Nelson Advanced Science

Organic Chemistry, Energetics, Kinetics and Equilibrium

revised edition

Brian Chapman • Alan Jarvis

Endorsed by

First published in 2000 by:
Thomas Nelson and Sons Ltd

This edition published in 2003 by:
Nelson Thornes Ltd
Delta Place
27 Bath Road
CHELTENHAM
GL53 7TH
United Kingdom

05 06 07 / 10 9 8 7 6 5 4 3

A catalogue record for this book is available from the British Library

ISBN 0 7487 7656 7

Illustrations by Hardlines and Wearset
Page make-up by Hardlines and Wearset

Printed in Croatia by Zrinski

Acknowledgements

Alan Thomas: 1.6b, 1.7, 3.3;
Andy Ross Photography: 3.4, 3.5;
Corbis: 3.7 (Patrick Ward), 3.10 (Bettmann);
Getty Images: 3.1 (Tony Stone Images, Arnulf Husmo), 6.2 (Tony Stone Images, Fred
George);
NT Pictures: 1.1, 3.2, 3.6, 3.9, 4.2;
Science Photolibrary: 3.8 (Dr Jeremy Burgess), 6.1 (Adam Hart-Davis)

Contents

Introduction

Introduction

This series has been written by Examiners and others involved directly with the development of the Edexcel Advanced Subsidiary (AS) and Advanced (A) GCE Chemistry specifications.

Organic Chemistry, Energetics, Kinetics and Equilibrium is one of four books in the Nelson Advanced Science (NAS) series developed by updating and reorganising the material from the Nelson Advanced Modular Science (AMS) books to align with the requirements of the specifications from September 2000. The books will also be useful for other AS and Advanced courses.

Organic Chemistry, Energetics, Kinetics and Equilibrium provides coverage of Unit 2 of the Edexcel specification. The book first looks at the importance of energy changes (expressed in terms of enthalpy) to a wide range of reactions and to the stability of compounds. A traditional, non-mechanistic approach is taken for the study of organic chemistry, with the examination of aliphatic compounds, their preparation and properties; it ends with a look at some commercial applications. The study of kinetics is entirely non-mathematical, the aim is to establish an appreciation of what controls rates rather than their measurement. The treatment of equilibrium is also qualitative. These lead to an understanding of the factors that must be considered when reversible reactions are used industrially.

Other resources in this series

NAS *Teachers' Guide for AS and A Chemistry* provides advice on safety and risk assessment, suggestions for practical work, key skills opportunities and answers to all the practice and assessment questions provided in *Structure, Bonding and Main Group Chemistry; Organic Chemistry, Energetics, Kinetics and Equilibrium; Periodicity, Quantitative Equilibria and Functional Group Chemistry;* and *Transition Metals, Quantitative Kinetics and Applied Organic Chemistry.*

NAS *Make the Grade in AS and A Chemistry* is a Revision Guide for students. It has been written to be used in conjunction with the other books in this series. It helps students to develop strategies for learning and revision, to check their knowledge and understanding and to practise the skills required for tackling assessment questions.

Features used in this book

The Nelson Advanced Science series contains particular features to help you understand and learn the information provided in the books, and to help you to apply the information to your coursework.

These are the features that you will find in the Nelson Advanced Science Chemistry series:

Text encapsulates the necessary study for the Unit. Important terms are indicated in **bold**.

5 Oxidation/reduction: an introduction

Introduction

Oxidation and reduction are found with all but four elements in the Periodic Table, not just with the transition metals, although they show these reactions to such an extent that they could be accused of self-indulgence.

When magnesium reacts with oxygen (Figure 5.1)

$$2Mg(s) + O_2(g) \rightarrow 2MgO(s)$$

the product contains Mg^{2+} and O^{2-} ions. Reaction with oxygen is pretty clearly oxidation. The reaction of magnesium with chlorine

$$Mg(s) + Cl_2(g) \rightarrow MgCl_2(s)$$

gives a compound with Mg^{2+} and Cl^- ions. In both cases the magnesium atom has lost electrons, so as far as the magnesium is concerned the reactions are the same. This idea is generalised into the definition of oxidation as loss of electrons. Reduction is therefore the gain of electrons. Since electrons don't vanish from the universe, oxidation and reduction occur together in **redox** reactions.

Oxidation numbers

For simple monatomic ions such as Fe^{2+} it's easy to see when they are oxidised (to Fe^{3+}) or reduced (to Fe). For ions such as NO_3^- or SO_3^{2-} which also undergo oxidation and reduction it is not always so easy to see what is happening in terms of electrons. To assist this, the idea of **oxidation number** or **oxidation state** is used. The two terms are usually used interchangeably, so that an atom may have a particular oxidation number or be in a particular oxidation state.

> **DEFINITION**
>
> **Oxidation** is electron loss.
> **Reduction** is electron gain.

> **MNEMONIC**
>
> **OIL RIG:**
>
> **o**xidation **is l**oss
>
> **r**eduction **is g**ain

Figure 5.1 The use of magnesium flares in photography being demonstrated at an early meeting of the British Association in Birmingham (1865).

Definition boxes in the margin highlight some important terms.

> **DEFINITION**
>
> **First ionisation energy:** the amount of energy required per mole to remove an electron from each atom in the gas phase to form a singly positive ion, that is
>
> $$M(g) \rightarrow M^+(g) + e^-$$
>
> **Second ionisation energy:** the energy per mole for the process
>
> $$M^+(g) \rightarrow M^{2+}(g) + e^-$$
>
> and so on for successive ionisation energies.

INTRODUCTION

Questions in the margin will give you the opportunity to apply the information presented in the adjacent text.

The **empirical formula** shows the ratio of atoms present in their lowest terms, i.e. smallest numbers. Any compound having one hydrogen atom for every carbon atom will have the empirical formula CH; calculation of the **molecular formula** will need extra information, since ethyne, C_2H_2, cyclobutadiene, C_4H_4, and benzene, C_6H_6, all have CH as their empirical formula. Empirical formulae are initially found by analysing a substance for each element as a percentage by mass.

QUESTION

Find the empirical formula of the compound containing C 22.02%, H 4.59%, Br 73.39% by mass.

Practice questions are provided at the end of each chapter. These will give you the opportunity to check your knowledge and understanding of topics from within the chapter.

Assessment questions are found at the end of the book. These are similar to the assessment questions for Advanced Subsidiary that you will encounter in your Unit Tests (exams) and they will help you to develop the skills required for these types of questions.

Questions

1 Plot on a (small) graph the first ionisation energies of the elements from sodium to argon, and account for the shape obtained.

2 Use data from a data book to plot a graph of atomic radius vs atomic number for the elements of Periods 2 and 3 (Li to Ar). Account for the difference in the graphs between Groups 2 and 3.

3 Sketch the structures of:
   (a) the giant covalent lattice of silicon
   (b) the molecule P_4
   (c) the molecule S_8.

4 Silicon has no compounds in which the silicon atom forms double bonds with other elements. Phosphorus, by contrast, does form double bonds with other elements. Suggest why silicon and phosphorus are different in this respect.

About the Authors

Brian Chapman was formerly Head of Science at Hardenhuish Comprehensive School until he retired. He is a retired Chief Examiner in Chemistry for Edexcel.

The late **Alan Jarvis** was former Head of Chemistry at Stoke-on-Trent Sixth Form and was Chief Examiner in Chemistry for Edexcel.

Acknowledgements

The authors and publisher would like to thank Geoff Barraclough for his work as Series Editor for the original series of four NAMS books, from which the new suite of NAS books was developed. Thanks also to Ray Vincent, Assessment Leader at Edexcel, who checked both the text and the assessment questions for the NAS student books.

Energy – The driving force of life

Life on Earth has always depended on the availability of energy, either directly or indirectly, from the Sun. Modern industrial societies, however, are even more dependent on the ready availability of energy in various forms. Fuels such as natural gas, oil and coal are burned in enormous quantities in our homes, our transport and in industry. These fuels are used directly as sources of heat for cooking, driving engines, etc. and are also converted into other forms of energy, such as electricity, in order to provide light and to drive other kinds of machinery. Within our own bodies, energy is obtained in a variety of ways. In muscles, for example, energy is obtained from the hydrolysis of large phosphate-containing molecules (see *Respiration and Coordination*, Adds et al., Nelson 2001).

Fig. 1.1 Heavy industry, such as the steel industry, uses huge amounts of energy.

All these energy-yielding processes are chemical reactions and this is reason enough for us to want to enquire into the relationship between chemical reactions and energy. In addition, however, a study of energy changes can lead to a better understanding of many of the fundamental processes in chemistry.

This chapter is about the relationship between chemical reactions and heat changes, although chemical reactions can produce energy in many different forms such as heat, light, electricity, etc. The relationship between chemical reactions and heat changes is often referred to as **thermochemistry** which is itself part of a larger study known as **thermodynamics**.

Signs, symbols and terminology

Enthalpy change

Heat changes have different values depending on the conditions under which they are measured, in particular whether they are measured at constant volume or constant pressure. The latter is more appropriate at this level since most reactions will be carried out in open containers, that is effectively at atmospheric (constant) pressure. The heat change measured at constant pressure is known as the enthalpy change and is given the symbol ΔH.

An enthalpy change always refers to a specified quantity of material; this is normally 1 mole or the molar amounts associated with an accompanying equation. The enthalpy change is then correctly given the symbol ΔH_m, but the 'm' is often omitted since enthalpy changes seldom refer to anything other than molar amounts. Materials are thought of as having a certain enthalpy content (H) – rather like a bank balance – and the enthalpy change measures the increase in enthalpy. In endothermic reactions, heat is absorbed and the products have more enthalpy than the reactants. In exothermic reactions, heat is evolved and this causes a reduction in enthalpy content.

ENERGY – THE DRIVING FORCE OF LIFE

Standard conditions

Standard conditions refer to the internationally agreed conditions under which an enthalpy change should be measured if it is to be called a standard enthalpy change (represented by the symbol $\Delta H^{\ominus}$). These conditions are a temperature of 298 K and a pressure of 1 bar or 1.00×10^5 Pa (approximately 1 standard atmosphere).

Units

Heat is a form of energy and both are measured in the same units. The unit used is that derived from the basic SI units, i.e. the **joule (J)**, although this is rather small for most chemical reactions and the **kilojoule (kJ)** is more frequently encountered. Another unit, known as the calorie, may be encountered occasionally, e.g. in food chemistry. This will not be used in this book but should it be met elsewhere, the conversion factor is 1 cal = 4.184 J.

Sign convention

Chemical reactions which liberate heat energy to the surroundings are known as **exothermic** reactions. Such reactions are accompanied by an increase in the temperature of the surroundings. Chemical reactions which absorb heat energy from the surroundings are known as **endothermic** reactions. Such reactions are accompanied by a decrease in the temperature of the surroundings.

With an exothermic process, it may seem strange that a process which 'gives out heat' (ΔH –ve) apparently becomes warmer; after all, if a hot block of metal 'gives out heat' it cools. If a few cm^3 of bench dilute hydrochloric acid is mixed with an equal volume of bench aqueous sodium hydroxide, the tube and solution feel very warm. The exothermic reaction

$$H^+(aq) + OH^-(aq) \rightarrow H_2O(l)$$

liberates heat which first warms the water and the tube (actually the most immediate surroundings of the reacting ions), and when the original temperature conditions are restored, that heat passes to the surrounding room (as it did from the hot block).

The sign convention used to distinguish between these two types of reaction is that a negative value for ΔH indicates an exothermic change, and a positive value for ΔH indicates an endothermic change.

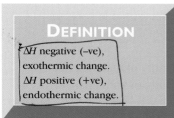
The sign is applied to the value of ΔH but does not indicate its magnitude. Thus a value of –400 kJ means 400 kJ of heat is evolved and is bigger than say –200 kJ which indicates that only 200 kJ of heat is evolved. Thus the enthalpy change for a reaction can be represented as follows:

$$C(s) + O_2(g) \rightarrow CO_2(g) \quad \Delta H^{\ominus} = -393 \text{ kJ mol}^{-1}$$

This is interpreted to mean that when one mole of solid carbon reacts with one mole of gaseous oxygen, gaseous carbon dioxide is formed and 393 kJ of heat are evolved when measured under standard conditions.

In writing equations such as this, it is essential that state symbols are shown since a change of physical state is accompanied by an enthalpy change. Indeed in some cases it is necessary to be even more specific about the actual substances used. For example the reaction above should more properly be written as:

$$C(s, graphite) + O_2(g) \rightarrow CO_2(g) \qquad \Delta H^{\ominus} = -393 \text{ kJ mol}^{-1}$$

as the values for the other allotropes of carbon are different, e.g.

$$C(s, diamond) + O_2(g) \rightarrow CO_2(g) \qquad \Delta H^{\ominus} = -395 \text{ kJ mol}^{-1}$$

A negative sign for ΔH thus indicates that energy has been *lost* from the substance involved and transferred to the surroundings.

INFORMATION

If you are asked for a definition of **standard** enthalpy change do not forget to add 'at 298 K and 100 kPa' to your definition. To add 'under standard conditions' may not satisfy the examiner.

Enthalpy diagrams

Chemical reactions involve the breaking of some bonds and the making of others. The reagents are regarded as having a certain amount of energy (or enthalpy) within them, as are the products. The enthalpy change of a reaction represents the difference in enthalpy between the reactants and products. This can be represented on an enthalpy diagram which has an increasing scale of energy as the vertical axis. This can be an actual scale with units specified or simply an arbitrary scale with unspecified units. E.g. for any reaction:

Reactants (R) $\rightarrow$ Products (P)

the enthalpy diagram would be as shown in Figure 1.2 or Figure 1.3.

In all cases,

$$\Delta H = H_{Products} - H_{Reactants}$$

Thus the value of ΔH is negative for the exothermic reaction and positive for the endothermic reaction.

Enthalpy diagrams give us a good indication of whether the reactants are less stable or more stable than the products. Thus in Figure 1.4, liquid water clearly has less enthalpy than gaseous hydrogen and oxygen, and is more stable than they are. However, you can mix hydrogen and oxygen and leave them 'for ever' and they may not react (see p.67). Stability judged in this way should be qualified as **thermodynamic**, e.g. liquid water is more thermodynamically stable than gaseous hydrogen and oxygen.

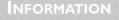

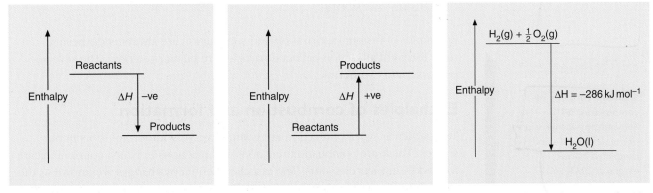

Fig. 1.2 Enthalpy diagram for an exothermic reaction.

Fig. 1.3 Enthalpy diagram for an endothermic reaction.

Fig. 1.4 The exothermic formation of water.

During much of the nineteenth century, it was thought that an exothermic change would always 'go' in the exothermic direction – and of course it usually does or they would not have come to that conclusion! Nevertheless, the conclusion is not valid. Add a little sodium nitrite to some water in a test tube and shake it. As the salt dissolves the mixture becomes very cold – the change you can see is endothermic and yet it happens remarkably easily.

For a true measure of thermodynamic stability, we need to refer to the 'free energy' difference between the products and reactants, ΔG. For many reactions in which ΔH is numerically large it is very similar to ΔG, except at very high temperatures; thus ΔH is a guide to stability but one which must be treated with caution. A reaction that occurs spontaneously goes in the direction of decreasing free energy, i.e. ΔG is –ve; ΔH will often be –ve too. (A discussion of the exact nature of free energy exceeds the scope of this book.)

Changes of state

As stated previously, it is important to specify the physical state of the substances involved when writing equations to represent an enthalpy change. This is because any change in physical state is itself accompanied by an enthalpy change. The changes of state are named as in Figure 1.5.

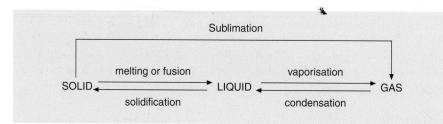

Fig. 1.5 Changes of state

Hence the **enthalpy of fusion** of H_2O would be the enthalpy change for the process

$$H_2O(s) \rightarrow H_2O(l)$$

and the **enthalpy of vaporisation** would be the enthalpy change for the process

$$H_2O(l) \rightarrow H_2O(g)$$

Enthalpies of fusion, vaporisation and sublimation are always endothermic since forces of attraction between particles are being overcome. The reverse processes are exothermic.

Enthalpies of combustion and formation

Two particular types of reaction are going to be very important to us in our studies. These are combustion reactions and reactions in which compounds are formed from their elements. The associated enthalpy changes are known as the **enthalpy of combustion** and the **enthalpy of formation** respectively.

DEFINITION

Standard enthalpy of combustion, $\Delta H^{\ominus}_c$
The enthalpy change when one mole of a substance is completely burned in oxygen at 298 K and 100 kPa.

These are defined as follows:

- The **standard enthalpy of combustion**, $\Delta H_c^{\ominus}$, is the enthalpy change when one mole of a substance is completely burned in oxygen, at 298 K and 100 kPa.
- The **standard enthalpy of formation**, $\Delta H_f^{\ominus}$, is the enthalpy change when one mole of a compound is formed from its elements in their standard states, at 298 K and 100 kPa.

The enthalpy of formation of any **element** in its standard state is zero.

The standard state of an element is generally the most stable form of the element under standard conditions – the form in which we would expect to find it. Thus $H_2(g)$, $O_2(g)$ and $Cu(s)$ are standard states. $H(g)$, $O_3(g)$ and $Cu(l)$ exist, but are not standard states.

Note that the combustion will not take place under standard conditions but the measurement of ΔH must be made when the conditions at the start and at the end of the reaction are standard (or they must be corrected appropriately).

Enthalpy of combustion

The enthalpy of combustion is obviously associated with the process of burning or combustion in oxygen. The process is always exothermic. In writing equations, the product(s) of complete combustion need to be known, for example, when carbon is burned completely the product is carbon dioxide and not carbon monoxide, while for sulphur it is sulphur dioxide and not sulphur trioxide. Hence, the equation which represents the enthalpy of combustion of sulphur (rhombic) is:

$$S(s, \text{rhombic}) + O_2(g) \rightarrow SO_2(g) \quad \Delta H^{\ominus} = -296.9 \, \text{kJ mol}^{-1}$$

The value for monoclinic sulphur would be different:

$$S(s, \text{monoclinic}) + O_2(g) \rightarrow SO_2(g) \quad \Delta H^{\ominus} = -297.2 \, \text{kJ mol}^{-1}$$

Most organic compounds burn in oxygen, with hydrocarbons (which contain carbon and hydrogen only) and carbohydrates (which contain carbon, hydrogen and oxygen) being particularly important. The products of combustion for these compounds are always carbon dioxide and water.

$$CH_4(g) + 2O_2(g) \rightarrow CO_2(g) + 2H_2O(g) \quad \Delta H^{\ominus} = -882 \, \text{kJ mol}^{-1}$$

Sucrose produces even more energy per mole:

$$C_{12}H_{22}O_{11}(s) + 12O_2(g) \rightarrow 12CO_2(g) + 11H_2O(g)$$
$$\Delta H^{\ominus} = -5644 \, \text{kJ mol}^{-1}$$

In reactions such as these, the water may be produced in the liquid or the gaseous state, depending on the temperature at which the reaction is carried out. If liquid water is produced, the values of the enthalpy changes will then be slightly different from those quoted above.

Calorific values

Most fuels and foodstuffs now display a variety of information on the packaging, which includes the energy value of the fuel or food. These are based on enthalpies of combustion since combustion and the processes that the fuel or food undergoes in the body give rise to the same product. There is no reference to moles in the information given, since the fuel or food is not sold by moles but by weight or volume, nor is there any reference to a minus sign in the values although all are exothermic. British Gas, for example, quote a 'calorific' value of $38.4\,MJ\,m^{-3}$ (megajoules per cubic metre; $1\,MJ = 10^6\,J$) and the cost to the consumer is calculated from the number of cubic metres used as measured by a gas meter installed on the premises.

Calorific values of foods are of vital importance for people who have to control their energy intake for dietary reasons. The values are displayed on the packaging as the number of units of energy which a given mass of food will provide on consumption. Some typical values are shown in Table 1.1, although actual values will depend on the exact brand used.

Table 1.1 *The calorific value of some common foods, per 100g*

Food	kJ	kcal
Potatoes (boiled)	342	82
Potato crisps	2350	562
White bread	1068	255
Brown bread	920	220
Beef	940	225
Butter	3031	724
Sunflower oil spread	2610	624

QUESTION

The calorific value of brown bread is not much different from that of white bread, but there is a huge difference between the calorific values of potato crisps and boiled potatoes. Why? (Try burning a sample of each using a pair of tongs.)

The energy value of foods eaten must equal the energy expenditure of our everyday activities. If we eat more than this, the body stores the excess in the form of fat, which is available for future use, but which is likely to cause numerous health problems. A 'calorie-controlled' diet involves eating foods with a lower energy value than the amount of energy expended. This forces the body to use up some of its reserves of fat in order to provide the additional energy required, and so weight is lost.

The relationship between enthalpy of formation and enthalpy of combustion

The enthalpy of formation refers to the formation of a compound from its elements, as defined earlier. When the compound formed is an oxide, this often involves the same equation as the combustion of the element. Consequently the enthalpy of formation of the oxide is often the same as the enthalpy of combustion of the element.

For example, the enthalpy of combustion of graphite is the same as the enthalpy of formation of carbon dioxide ($-393\,kJ\,mol^{-1}$), since both are in fact the enthalpy change for the reaction

$$C(s, graphite) + O_2(g) \rightarrow CO_2(g) \qquad \Delta H^\ominus = -393\,kJ\,mol^{-1}$$

This cannot be universally applied, however. For example, the equation

$$C(s, \text{diamond}) + O_2(g) \rightarrow CO_2(g) \quad \Delta H^\ominus = -395 \, \text{kJ mol}^{-1}$$

represents the enthalpy of combustion of diamond but does not represent the enthalpy of formation of carbon dioxide since the carbon is not in its standard state. Graphite is the standard state for carbon since it is thermodynamically more stable than diamond at 298 K, as the enthalpy changes above show. Similarly, the enthalpy of formation of lithium oxide is represented by the following equation

$$2\text{Li}(s) + \tfrac{1}{2}O_2(g) \rightarrow \text{Li}_2O(s) \quad \Delta H^\ominus = -596 \, \text{kJ mol}^{-1}$$

but this is not the enthalpy of combustion of lithium since two moles of lithium are involved in the equation.

Enthalpy of neutralisation

When an acid and an alkali neutralise each other the reaction is exothermic. The enthalpy term is a little more difficult to define than ΔH_c or ΔH_f, which invariably refer to the combustion or formation of 1 mole of a compound. You have already learned to represent neutralisation by the ionic equation

$$H^+(aq) + OH^-(aq) \rightarrow H_2O(l)$$

Although this equation is not always appropriate at GCE Advanced level and beyond (see Unit 4), in order to produce an enthalpy term which is comparable between different acids and alkalis it is better to relate ΔH to the quantities associated with the above equation rather than to 1 mole of an acid which may give rise to 1, 2 or 3 moles of hydrogen ions, i.e. it is better to refer to the enthalpy change when a specified acid reacts with a specified base at a specified concentration (usually 1 mol dm^{-3}) to produce 1 mole of water or its equivalent (see below).

Even then, ambiguity can still arise. Does the neutralisation of sulphuric acid refer to $\tfrac{1}{2}$ mol of H_2SO_4 (the usual case)

$$NaOH(aq) + \tfrac{1}{2}H_2SO_4(aq) \rightarrow \tfrac{1}{2}Na_2SO_4(aq) + H_2O(l)$$

or to the first of two stages?

$$NaOH(aq) + H_2SO_4(aq) \rightarrow NaHSO_4(aq) + H_2O(l)$$

It is thus very important to make clear (by writing an associated equation), to what amount of acid or alkali ΔH_{neut} refers.

If the earlier equation

$$H^+(aq) + OH^-(aq) \rightarrow H_2O(l)$$

truly represented every neutralisation, then every enthalpy of neutralisation, referred to 1 mol of water, should be the same. The enthalpy of formation of water from its hydrated ions is –57 kJ mol^{-1}. The neutralisations shown in Table 1.2 are clearly well represented by this equation.

> **DEFINITION**
>
> The [*standard*] enthalpy of neutralisation is the heat change when aqueous solutions of an acid and a base of specified concentration (1.00 mol dm^{-3}) react in the quantities associated with the transfer of one mole of protons, usually to give one mole of water, according to a specified equation [*at 298 K and 100 kPa*].

However, when we consider ethanoic acid (CH_3COOH) or ammonia (NH_3), only a small percentage (typically 2%) of the solution is ionised, and the bulk of each of these compounds is present as unionised molecules. The dominant reactions are:

$$CH_3COOH(aq) + OH^-(aq) \rightarrow H_2O(l) + CH_3COO^-(aq)$$

and

$$NH_3(aq) + H^+(aq) \rightarrow NH_4^+(aq)$$

The enthalpies observed are:

Acid	Alkali	ΔH/kJ mol^{-1}
Ethanoic	NaOH	−56.0
HCl	Ammonia	−51.3

Ethanoic acid and ammonia will later be referred to as a *weak* acid and a *weak* base. Anticipating Unit 4 for a moment, we can see that the neutralisations above all refer to the transfer of one mole of protons (H^+) from an acid (HCl, HNO_3 or $\frac{1}{2}H_2SO_4$) to a base (OH^- or NH_3). It is thus better to define enthalpy of neutralisation, in general, in terms of the transfer of one mole of protons since this covers the examples, that you will meet later, of weak acids and bases where water is not necessarily a product.

Experimental measurement of enthalpy changes

Heat and temperature

The difference between these two terms must be made clear if the measurement techniques are to be understood. Temperature does not represent the amount of heat energy of a substance but merely the degree of hotness on some arbitrary scale. It is essentially a measure of the kinetic energy of the molecules or particles present in the substance and is independent of the amount of substance present. Heat, on the other hand, is a measure of the total energy in a substance and does depend on the amount of substance present. Thus on comparing a large bucket of water at a temperature of 50 °C with say 250 cm^3 of water in a beaker at 50 °C, you can see that both have the same temperature but the bucket of water would contain much more heat energy in total.

The amount of heat required to raise the temperature of 1 g of a substance by 1 K is called the **specific heat capacity** (c) and the value differs from one substance to another. Water, for example, has a value of about 4.2 J g^{-1} K^{-1} while that for ethanol is about 2.4 J g^{-1} K^{-1}. The value for water is exceptionally high and hence water is able to store much more heat per unit mass than most other liquids.

Heat is always transferred from a hot substance to a cold substance and this will produce a change of temperature which can be measured. Thus although there is no instrument which measures heat directly, the amount of heat transferred from a hot substance to a cold substance can be calculated if the mass of one substance, its specific heat capacity and the temperature change, are known.

Table 1.2 Some enthalpies of neutralisation

Acid	Alkali	ΔH/kJ mol^{-1}
HNO_3	NaOH	−57.2
HCl	LiOH	−57.3
HCl	NaOH	−57.2
HCl	KOH	−58.2

INFORMATION

Ac is a useful abbreviation for ethanoyl (CH_3CO-), which has been retained from the days when it was called acetyl.

DEFINITION

Specific heat capacity The amount of heat required to raise the temperature of 1 g of substance by 1 K.

QUESTION

Approximately how much heat escapes from 150 cm^3 of tea when it cools from 80 °C to 50 °C?

The heat transferred is then given by the expression:

heat transfer = mass × specific heat capacity × temperature change

heat transfer = $m \times c \times \Delta T$

Consistent units must of course be used and these will depend on the units given for c. These are usually $J\,g^{-1}\,°C^{-1}$ in which case m must be in grams. There is no need to change temperatures measured in °C to K since it is a temperature **difference** which is measured, and a change of 1 °C is the same as a change of 1 K. The heat transfer will then be in joules. Measurement of heat therefore depends on transferring the heat to a known mass of another substance (usually water) and measuring the temperature rise. This is the basis of **calorimetry** and the apparatus used is called a **calorimeter**.

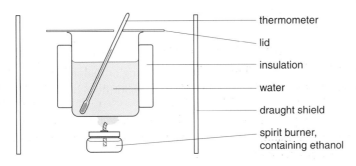

- thermometer
- lid
- insulation
- water
- draught shield
- spirit burner, containing ethanol

Fig. 1.6a A simple form of calorimeter.

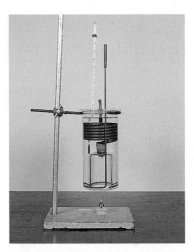

Fig 1.6b A Thiemann calorimeter for measuring enthalpies of combustion.

Enthalpy of combustion

A very simple form of calorimeter which could be used to measure the enthalpy of combustion of a liquid such as ethanol is shown in Figure 1.6a. A known mass of water (150 g) is placed in a glass beaker or a metal can and its temperature noted (23 °C). A spirit burner containing ethanol is weighed and the burner is placed under the beaker and lit. The water in the beaker is stirred and after a certain time the final temperature of the water is noted (43 °C), the spirit burner is extinguished and it is re-weighed. The difference in mass of the spirit burner initially and at the end of the experiment gives the mass of ethanol burned (0.90 g). Thus:

heat gained by water in calorimeter $= m \times c \times \Delta T$
$$= 150\ g \times 4.2\ J\,g^{-1}\,°C^{-1} \times 20\ °C$$
$$= 12\ 600\ J$$

Hence the heat produced by burning 0.90 g of ethanol = 12.6 kJ

M_r (C_2H_5OH) $=$ 46 g mol^{-1}

Ethanol used $= \dfrac{0.90\ g}{46\ g\ mol^{-1}} = 0.01956$ moles

So heat produced by burning **1 mol** of ethanol $= 12.6 \times \dfrac{46}{0.90} = 644$ kJ

$\Delta H_c = -640$ kJ mol^{-1}

QUESTION

In a similar experiment 1.0 g of propanol raised the temperature of the same amount of water by 25.1 °C. What result does this give for ΔH_c (propanol)? What does previous experience suggest might be a more accurate figure?

The last figure is reduced to 2 significant figures (**64**0 or, unambiguously, $0.\mathbf{64} \times 10^3$, **not** $0.\mathbf{640} \times 10^3$), because most of the data is only expressed to 2 significant figures, e.g. **23** °C, **43** °C or 0.**90** g).

The negative sign is inserted since the reaction must be exothermic, a temperature rise having occurred.

This value is numerically much less than the accepted value of -1371 kJ mol^{-1}. Some of the reasons for this are:
- heat is still lost from the calorimeter to the surroundings despite the use of a lid and insulation of the sides.
- some heat goes into the calorimeter instead of the water. This could be allowed for if the mass of the calorimeter and its specific heat capacity were known, or by calibrating the apparatus appropriately.
- incomplete combustion of the ethanol due to an inadequate supply of oxygen, leading to the formation of products such as carbon monoxide or even carbon (as indicated by a deposit of soot on the bottom of the calorimeter).

More accurate methods are available for measuring enthalpies of combustion, e.g. using a bomb calorimeter, but these are beyond the scope of this book and will be found in textbooks of physical chemistry.

Measuring enthalpy changes for reactions in solution

For reactions which take place in solution, the heat is generated (or absorbed) within the solutions themselves and it is simply a matter of keeping the heat within the solution. A calorimeter is therefore selected which is a very good insulator in order to reduce the heat escaping to the surroundings or being absorbed from the surroundings. A vacuum flask is very good in this respect, although in school laboratories a beaker made from expanded polystyrene is more frequently used. This material has a very low specific heat capacity and hence absorbs very little heat itself. The apparatus is shown in Figure 1.7 and good results can be obtained from this relatively simple apparatus. For example, the molar enthalpy change ΔH_{neut} for the reaction:

$$HCl(aq) + NaOH(aq) \rightarrow NaCl(aq) + H_2O(l)$$

can be measured as follows.

Place 50 cm^3 of a 1.0 mol dm^{-3} solution of hydrochloric acid in an insulated polystyrene cup and note its temperature (Figure 1.8). Add 50 cm^3 of a 1.1 mol dm^{-3} sodium hydroxide solution (an excess to ensure complete reaction) which is at the same temperature. Stir continuously and note the maximum temperature reached. The temperature rise will be about 6.5 °C. The total volume of solution will be 100 cm^3 which we will assume to be 100 g.

> **INFORMATION**
>
> In approximate experiments the density of aqueous solutions is often taken to be 1.0 g cm^{-3}.

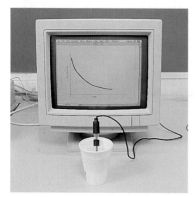

Fig. 1.7 Polystyrene cup calorimeter attached via a datalogger to a computer, producing a graph which shows change in temperature over time.

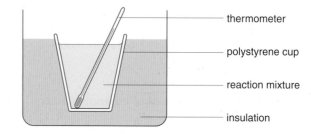

thermometer

polystyrene cup

reaction mixture

insulation

Fig 1.8 Measuring enthalpy changes for reactions in solution.

Hence assuming no heat losses to the surroundings and assuming that the specific heat capacity of the solution is the same as that of water $(4.2 \, \text{J} \, \text{g}^{-1} \, {}^\circ\text{C}^{-1})$, then:

$$\text{Heat absorbed by solution} = m \times c \times \Delta T$$

$$= 100 \, \text{g} \times 4.2 \, \text{J} \, \text{g}^{-1} \, {}^\circ\text{C}^{-1} \times 6.5 \, {}^\circ\text{C}$$

$$= 2730 \, \text{J}$$

$$\text{Amount of acid} = \frac{50}{1000} \times 1.0 = 0.050 \, \text{mol}$$

$$\text{Heat given by 1 mole of acid} = \frac{2730 \, \text{J}}{0.050 \, \text{mol}} = 54 \, 600 \, \text{J} \, \text{mol}^{-1}$$

$$\therefore \Delta H_{\text{neut}} = -55 \, \text{kJ} \, \text{mol}^{-1}$$

The negative sign is inserted since the reaction is obviously exothermic because a temperature rise was observed. The errors in this method arise mainly through the assumptions that the densities and specific heat capacities of the solutions are those of water. Without correction for this, such a method would be ill suited to showing, for example, the different behaviour of ammonia and sodium hydroxide solutions on neutralisation with hydrochloric acid.

Temperature corrections

As seen above, reasonably accurate results can be obtained using the simple apparatus described. The results are less accurate, however, if the reaction being performed is slower than the neutralisation reaction used above. This is because heat loss to the surroundings will increase if the reaction is slow because the heat will be lost over a longer period. This means that the temperature rise observed in the calorimeter is never as great as it should be. An allowance can be made for this by plotting a temperature–time graph. One reagent is placed in the polystyrene cup and its temperature noted at say 1 min intervals for say 4 min, stirring continuously. At a known time, say 4.5 min, the second reagent is added, stirring continuously, and the temperature noted more frequently until the maximum is reached. As the solution starts to cool the temperature is still recorded and stirring continued, for at least 5 min longer. A graph of temperature against time is then plotted. Graphs are given for an exothermic reaction (Figure 1.9) and for an endothermic reaction (Figure 1.10).

QUESTION

Why has the negative sign been included in this answer? Why is the answer not $-54.6 \, \text{kJ} \, \text{mol}^{-1}$?

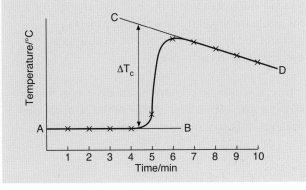

Fig. 1.9 A temperature correction curve for an exothermic reaction.

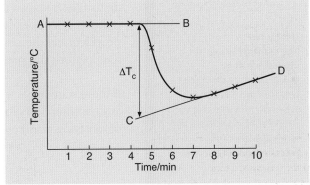

Fig 1.10 A temperature correction curve for an endothermic reaction.

The corrected temperature rise or fall is then obtained from the graph by the following method:
- draw the best straight line through the results for the first 4 min (A–B)
- draw the best straight line through the results after the peak, when the temperature is falling (or rising) (C–D)
- continue both lines to the time of mixing (4.5 min in this case) and take the difference between them at this time.

This is the **corrected temperature rise** (ΔT_c). The corrected final temperature is greater than the maximum temperature observed for an exothermic reaction or is less than the minimum observed for an endothermic reaction. The amount by which this differs from the maximum (or minimum) observed is dependent on the steepness of the line after the maximum (or minimum) and this is dependent on the heat loss (or gain) from the surroundings. Thus the greater the heat losses to the surroundings, the more the temperature is corrected. Other methods of correcting temperature changes are available but this is the simplest and it is often used at this level.

Mass and specific heat capacity corrections
It is easy enough to determine the masses of the solutions used: this is likely to be the most important correction. The exact specific heat capacity of the solutions is more difficult: the data is difficult to find and experimental methods introduce their own errors. In elementary work, you may have to settle for accepting this error.

Hess's law and the calculation of enthalpy changes

The enthalpy changes of many chemical reactions cannot be measured directly by the methods indicated in the previous section. In such cases it is possible to calculate the enthalpy change, given suitable alternative data. The calculations are based on an application of the First Law of Thermodynamics, one statement of which is: energy cannot be created or destroyed. It can only be converted from one form into another. Although this requires some modification in order to be absolutely true, it is a good working statement at this stage.

Two important deductions can be made from this. The first is that if the enthalpy change of a reaction is known, then the enthalpy change of the reverse reaction has the same value but with the sign changed. The second deduction is known as Hess's law, which states that the enthalpy change for a reaction is independent of the route by which the reaction is achieved, provided that the physical states of the reactants and products, and their temperature and pressure, are the same in each case.

If these statements were not true, it would be possible to create energy without any consumption of material. Desirable as this might be from the point of view of satisfying world energy demands, it is unfortunately impossible.

The first deduction is important since it allows us to know the enthalpy change for reactions which are very difficult or even impossible to measure. For example the enthalpy change for the reaction:

$$CO_2(g) \rightarrow C(s, diamond) + O_2(g)$$

cannot be measured directly but its value must be $+395\,kJ\,mol^{-1}$ since the enthalpy change of the reverse reaction can be measured and has a value of $-395\,kJ\,mol^{-1}$.

Hess's law is important since it allows us to calculate enthalpy changes for reactions when these cannot be measured experimentally. Examples of how this can be achieved are given below.

Calculations using Hess's law

There are several different techniques for the performance of these calculations but all are essentially the same in that they require application of Hess's law.

Method 1: Constructing an alternative route

Consider the conversion of reactants A into products B, a reaction for which it is impossible to measure ΔH directly. A can however be converted into B via two other compounds, C and D. This can be represented on the cycle shown in Figure 1.11. The calculation is then completed by application of Hess's law which states that the total enthalpy change for Route 1 = the total enthalpy change for Route 2.

$$\therefore \Delta H = \Delta H_1 + \Delta H_2 + \Delta H_3$$

Hence ΔH can be calculated if values are known for ΔH_1, ΔH_2 and ΔH_3.

Example 1

Calculate the enthalpy change for the reaction C(s, graphite) $\rightarrow$ C(s, diamond), given the following data:

$$C(s, graphite) + O_2(g) \rightarrow CO_2(g) \qquad \Delta H^{\ominus} = -393\,kJ\,mol^{-1}$$

$$C(s, diamond) + O_2(g) \rightarrow CO_2(g) \qquad \Delta H^{\ominus} = -395\,kJ\,mol^{-1}$$

The alternative route for converting graphite to diamond is fairly obvious:

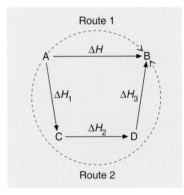

Fig. 1.11 A reaction cycle

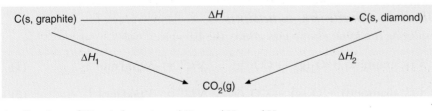

Application of Hess's law gives $\Delta H = \Delta H_1 - \Delta H_2$
From the data given $\Delta H_1 = -393\,kJ\,mol^{-1}$
$$\Delta H_2 = -395\,kJ\,mol^{-1}$$

Inserting appropriate values (including signs)

$$\Delta H = (-393) - (-395)$$

$$\Delta H = +2 \, \text{kJ mol}^{-1}$$

Errors will be avoided if signs are always inserted with the values of ΔH.

Simple problems of this sort could also be solved by putting the data onto an enthalpy diagram. If this is done correctly, the answer becomes obvious from the diagram. Thus the data for the combustion of graphite and diamond can be put on to an enthalpy diagram as shown in Figure 1.12.

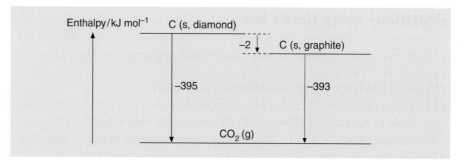

Fig. 1.12 An enthalpy diagram for graphite and diamond.

Graphite is therefore energetically more stable than diamond and it would take $+2 \, \text{kJ mol}^{-1}$ to convert one mole of graphite to one mole of diamond. Similarly $2 \, \text{kJ mol}^{-1}$ would be evolved if the reverse process was carried out.

In passing it might be noted that the conversion of graphite to diamond is not impossible, indeed many industrial diamonds are manufactured in this way. It is however an extremely difficult reaction to achieve, requiring very high temperatures and pressures. We might have expected such a difficult reaction to be endothermic but the value is so very small that this in itself is unlikely to make the reaction difficult to accomplish. A huge amount of energy must be put into the graphite to disrupt some of its bonds before the atoms can be rearranged into the diamond structure (with the release of all but $2 \, \text{kJ mol}^{-1}$ of the energy put in).

Method 2: Combining equations

Considering the same example as above, the data given was:

$$C(s, \text{graphite}) + O_2(g) \rightarrow CO_2(g) \qquad \Delta H_1^{\ominus} = -393 \, \text{kJ mol}^{-1} \qquad (1)$$

$$C(s, \text{diamond}) + O_2(g) \rightarrow CO_2(g) \qquad \Delta H_2^{\ominus} = -395 \, \text{kJ mol}^{-1} \qquad (2)$$

Reversing equation (2) gives:

$$CO_2(g) \rightarrow C(s, \text{diamond}) + O_2(g) \qquad \Delta H_3^{\ominus} = +395 \, \text{kJ mol}^{-1} \qquad (3)$$

Adding equations (1) and (3) gives:

$$C(s, \text{graphite}) + O_2(g) + CO_2(g) \rightarrow CO_2(g) + C(s, \text{diamond}) + O_2(g)$$

which, on cancelling common species, is the desired equation:

C(s, graphite) → C(s, diamond)

and, doing the same to the enthalpy values as was done to the equations gives:

$$\Delta H = \Delta H_1^\ominus + \Delta H_3^\ominus = (-393) + (+395) = +2\,\text{kJ}\,\text{mol}^{-1}$$

Both methods are equally valid and the choice is up to the individual. However, the following points should be checked carefully whichever method is adopted.

- Check that the sign of the ΔH value is correct for the equation used.
- If you multiply an equation (e.g. by 2) multiply the ΔH value as well.

Example 2
Given the following data:

$$2NO_2(g) \rightarrow 2NO(g) + O_2(g) \qquad \Delta H^\ominus = +109\,\text{kJ}\,\text{mol}^{-1}$$

$$\tfrac{1}{2}N_2(g) + \tfrac{1}{2}O_2(g) \rightarrow NO(g) \qquad \Delta H^\ominus = +90.0\,\text{kJ}\,\text{mol}^{-1}$$

$$N_2(g) + 2O_2(g) \rightarrow N_2O_4(g) \qquad \Delta H^\ominus = +8.0\,\text{kJ}\,\text{mol}^{-1}$$

Calculate the enthalpy change for the reaction:

$$N_2O_4(g) \rightarrow 2NO_2(g)$$

Using Method 1
The alternative cycle would be:

Applying Hess's law to this cycle gives:

$$\Delta H = \Delta H_1 + \Delta H_2 + \Delta H_3$$

$$\Delta H_1 = -8 + 2(+90) - 109$$
$$= +63\,\text{kJ}\,\text{mol}^{-1}$$

Using Method 2

Data:

$$\tfrac{1}{2}N_2(g) + \tfrac{1}{2}O_2(g) \rightarrow NO(g) \qquad \Delta H^\ominus = +90.0\,\text{kJ}\,\text{mol}^{-1} \qquad (4)$$

$$N_2(g) + 2O_2(g) \rightarrow N_2O_4(g) \qquad \Delta H^\ominus = +8.0\,\text{kJ}\,\text{mol}^{-1} \qquad (5)$$

$$2NO_2(g) \rightarrow 2NO(g) + O_2(g) \qquad \Delta H^\ominus = +109\,\text{kJ}\,\text{mol}^{-1} \qquad (6)$$

It is often difficult to see your way through a set of equations – especially if they are similar in appearance, like the *data* here. Try to pick species from your target equation, each of which occurs **once** only in the *data*. This is easy here because the target only contains two species. Thus, $N_2O_4(g)$ on the left hand side (LHS) of the target equation occurs **only** on the RHS (right hand side) of equation (5). Begin by reversing equation (5) (changing the sign of ΔH):

$$N_2O_4(g) \rightarrow N_2(g) + 2O_2(g) \qquad \Delta H^\ominus = -8.0 \text{ kJ mol}^{-1} \qquad (7)$$

Similarly, $NO_2(g)$ occurs **only** in equation (6), again on the wrong side (compared with the target) so reverse equation (6):

$$2N(g) + O_2(g) \rightarrow 2NO_2(g) \qquad \Delta H^\ominus = -109 \text{ kJ mol}^{-1} \qquad (8)$$

Adding these, and taking away one oxygen molecule which occurs on both sides:

$$N_2O_4(g) + \boxed{2NO(g) \rightarrow N_2(g) + O_2(g)} + 2NO_2(g)$$

$$\Delta H^\ominus = -109 + (-8.0) \text{ kJ mol}^{-1} = -117 \text{ kJ mol}^{-1} \qquad (9)$$

Now look to see how you can use the remaining equation(s), usually to get rid of accumulated but unnecessary material in your developing equation. Inspection shows that we can remove unwanted material (in the box) by doubling equation (4) (and the corresponding value of ΔH):

$$N_2(g) + O_2(g) \rightarrow 2NO(g) \qquad \Delta H^\ominus = +180 \text{ kJ mol}^{-1}$$

and adding the result to equation (9) to give the target equation:

$$N_2O_4(g) \rightarrow 2NO_2(g) \quad \Delta H^\ominus = +180 + (-117) \text{ kJ mol}^{-1} = +63 \text{ kJ mol}^{-1}$$

Method 3: Using standard enthalpies of formation

This is really just a variation of method 1 but has the advantage that it can be used in all situations provided that values of enthalpies of formation are given. **Starting with the necessary elements** in their standard states, you can make the products of a reaction either directly or via the 'Reactants':

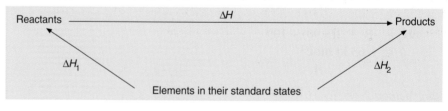

ΔH_2 = the sum of the enthalpies of formation of the products

ΔH_1 = the sum of the enthalpies of formation of the reactants

Hess's law tells us that the total enthalpy change is the same by either route:

Enthalpy change by direct route = ΔH_2

Enthalpy change by making Reactants first = $\Delta H_1 + \Delta H$

$\Delta H_1 + \Delta H = \Delta H_2$

$\Delta H = -\Delta H_1 + \Delta H_2$

This leads to the universally applicable formula:

$\Delta H = [\text{the sum of } \Delta H_f^\ominus (\text{products})] - [\text{the sum of } \Delta H_f^\ominus (\text{reactants})]$

In applying this formula it must be remembered that:
- the enthalpies of formation of **elements** are arbitrarily taken to be zero;
- if more than one molecule of a substance is involved, the enthalpy of formation must be multiplied appropriately;
- the sign of the enthalpy change must be inserted with the value;
- the formula does **not** apply to enthalpy (changes) of combustion.

Example 3
Calculate the standard enthalpy change for the combustion of ammonia in pure oxygen which occurs according to the equation:

$4NH_3(g) + 3O_2(g) \rightarrow 2N_2(g) + 6H_2O(g)$

given that the standard enthalpies of formation for $NH_3(g)$ and $H_2O(g)$ are -46.1 and $-242\,kJ\,mol^{-1}$, respectively.

Using the statement of Hess's law derived above:

$\Delta H^\ominus = \text{the sum of } \Delta H_f^\ominus(\text{products}) - \text{the sum of } \Delta H_f^\ominus(\text{reactants})$

$\therefore \Delta H^\ominus = \{6 \times (-242) + 2 \times 0\} - \{4 \times (-46) + 3 \times 0\}$

$\therefore \Delta H^\ominus = -1268\,kJ\,mol^{-1}$

Note that the units are **$kJ\,mol^{-1}$** despite the fact that there are 4 mol of ammonia in the equation and 6 mol of water. The units in this case refer to 1 mol of equation as written. If you were asked to calculate the standard enthalpy (change) of combustion of ammonia then

$\Delta H_c^\ominus(NH_3) = \dfrac{-1268}{4}\,kJ\,mol^{-1}$

$= -317\,kJ\,mol^{-1}$

This problem could be equally well answered by application of methods 1 and 2.

Average bond enthalpies

Covalent molecules can be broken up into the atoms of which they are composed by supplying energy in some form such as heat, light etc. and the enthalpy change for such a reaction can be measured. This is known as the **enthalpy of**

DEFINITION
Enthalpy of Dissociation
The enthalpy change when one mole of a gaseous substance is broken up into free gaseous atoms.

dissociation which is defined as the enthalpy change when one mole of a gaseous substance is broken up into free gaseous atoms. For example the enthalpy of dissociation of hydrogen is the enthalpy change for the reaction:

$$H_2(g) \rightarrow H(g) + H(g) \qquad \Delta H^\ominus = +432 \, kJ \, mol^{-1}$$

Since the atoms formed are in the gaseous state then the covalent bonds holding the atoms together in the molecule can be considered to have been completely broken. Hence this enthalpy of dissociation is a measure of the strength of the covalent bonds in one mole of molecules of H_2 and is also referred to as the **bond enthalpy** of hydrogen, represented by $E(H–H) = +432 \, kJ \, mol^{-1}$. The values are always positive.

For polyatomic molecules such as methane (CH_4), the enthalpy of dissociation would be the enthalpy change for the reaction:

$$CH_4(g) \rightarrow C(g) + 4H(g) \quad \Delta H^\ominus = +1664 \, kJ \, mol^{-1}$$

In this reaction, four covalent carbon–hydrogen bonds are broken and it would seem reasonable to assume therefore that the amount of energy required to break one such bond would be $1664/4 = 416 \, kJ \, mol^{-1}$. This is only an average or mean value however and is known as the **average bond enthalpy** represented by $E(C–H) = +416 \, kJ \, mol^{-1}$.

The specific bond enthalpies for the four C–H bonds are in fact quite different from one another, e.g.

$$CH_4(g) \rightarrow CH_3(g) + H(g) \quad \Delta H^\ominus = +427 \, kJ \, mol^{-1}$$

$$CH_3(g) \rightarrow CH_2(g) + H(g) \quad \Delta H^\ominus = +371 \, kJ \, mol^{-1}$$

The reasons for this are beyond the scope of this book. We shall concern ourselves with only the average bond enthalpies.

Average bond enthalpies and enthalpy of reaction

Average bond enthalpies can be used to calculate the enthalpy change for a reaction. This is done by assuming that an alternative route for all reactions can be achieved theoretically via the gaseous atoms of the elements involved in the compounds. It is therefore a specific application of Hess's law to the cycle:

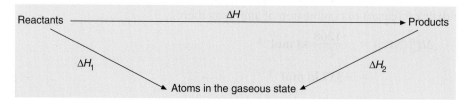

ΔH_1 = sum of the average bond enthalpies of the reactants

ΔH_2 = sum of the average bond enthalpies of the products

Applying Hess's Law gives:

$$\Delta H = \Delta H_1 - \Delta H_2$$

This leads to the universally applicable formula:

ΔH = [the sum of the average bond enthalpies of the reactants]

– [the sum of the average bond enthalpies of the products]

The method does require some knowledge of the bonding present in the particular molecules.

Example

Calculate the enthalpy change for the reaction:

$$CH_4(g) + Cl_2(g) \rightarrow CH_3Cl(g) + HCl(g)$$

given the average bond enthalpies in $kJ\,mol^{-1}$:

$$E(C–H) = 412;\ E(C–Cl) = 338;\ E(Cl–Cl) = 242;\ E(H–Cl) = 431$$

$$\therefore \Delta H^{\ominus} = [4 \times E(C–H) + E(Cl–Cl)] – [3 \times E(C–H) + E(C–Cl) + E(H–Cl)]$$
$$= [(4 \times 412) + 242] – [(3 \times 412) + 338 + 431]$$
$$= [1890] – [2005] = -115\,kJ\,mol^{-1}$$

> **QUESTION**
>
> Calculate the enthalpy change for the next stage of substitution to give CH_2Cl_2.

Alternative method

This involves a simple accounting procedure:

Energy absorbed (bonds broken) = $(4 \times 412) + 242 = +1890\,kJ$

Energy evolved (bonds formed) = $(3 \times 412) + 338 + 431 = -2005\,kJ$

Total enthalpy change = energy absorbed + energy evolved
$$= 1890 - 2005 = -115\,kJ\,mol^{-1}$$

The arithmetic can be simplified if it is realised that three of the C–H bonds broken in CH_4 are re-formed in CH_3Cl. Hence in total only one C–H bond is broken together with one Cl–Cl bond and one C–Cl bond is formed together with one H–Cl bond. Hence:

$$\Delta H^{\ominus} = [E(C–H) + E(Cl–Cl)] – [E(C–Cl) + E(H–Cl)]$$
$$= [412 + 242] – [338 + 431]$$
$$= 654 - 769 = -115\,kJ\,mol^{-1}$$

You must remember that such a calculation is *approximate*; there are two reasons for this.

The breaking of this particular C–H bond requires $427\,kJ\,mol^{-1}$ (see p.18) not $412\,kJ\,mol^{-1}$, but the formation of the C–Cl bond is probably in error by a similar amount and the errors will roughly cancel.

Another error that occasionally appears in some calculations using bond energies, arises because bond energies are concerned with *disrupting* molecules – not *separating* them from one another. They are based on gaseous molecules forming gaseous atoms. Suppose that you calculate $\Delta H_f(H_2O)$, i.e. the enthalpy change for the reaction

$$H_2(g) + \tfrac{1}{2}O_2(g) \rightarrow H_2O(l)$$

using the bond energies $E(\text{H–H})$ [in hydrogen gas] $= 436$ kJ mol^{-1}; $E(\text{O=O})$ [in oxygen gas] $= 498$ kJ mol^{-1} ; $E(\text{O–H}) = 464$ kJ mol^{-1} :

$$\Delta H = E(\text{H–H}) + \tfrac{1}{2}E(\text{O=O}) - 2E(\text{O–H})$$
$$= 436 + 249 - 928 = -243 \text{ kJ mol}^{-1}$$

If you look up $\Delta H_f(\text{H}_2\text{O})$, you will find the figure -286 kJ mol^{-1}! Rather a large error? Yes, but you will have made it!!! You will have calculated the enthalpy of formation of gaseous water – remember Hess's law and its insistence on the same state. You have not finished your calculation – you have still to calculate one more step. The enthalpy change for

$$\text{H}_2\text{O(g)} \rightarrow \text{H}_2\text{O(l)}; \ \Delta H = -44 \text{ kJ mol}^{-1}$$

which gives a final value for ΔH of $-243 + (-44) = -287$ kJ mol^{-1} – an excellent agreement.

Questions

1 (a) Define (i) enthalpy of formation;
         (ii) enthalpy of combustion.

   (b) When 12.00 g each of carbon and hydrogen are completely burned in oxygen, 393.5 and 1715.4 kJ are evolved respectively. Calculate the enthalpies of combustion of carbon and hydrogen.

2 (a) State Hess's law and give two examples to illustrate its usefulness in chemistry. No numerical data are required.

   (b) Calculate the enthalpy change for the following reaction, using the data below:

$$\text{P}_4\text{O}_{10}(s) + 6\text{H}_2\text{O(l)} \rightarrow 4\text{H}_3\text{PO}_4(s)$$

   Data: The enthalpies of formation of $\text{P}_4\text{O}_{10}(s)$, $\text{H}_2\text{O(l)}$ and $\text{H}_3\text{PO}_4(s)$ are -2984, -285.9 and -1279 kJ mol^{-1}, respectively.

3 Calculate the enthalpy of formation of butane $(\text{C}_4\text{H}_{10})$ using the following data:

   Enthalpy of combustion of graphite $= -393.5$ kJ mol^{-1}

   Enthalpy of combustion of hydrogen $= -285.9$ kJ mol^{-1}

   Enthalpy of combustion of butane $= -2877.1$ kJ mol^{-1}

4 (a) How is the enthalpy of formation of a substance connected to the thermodynamic stability of the substance?

   (b) Sulphur has two allotropes, rhombic and monoclinic, both of which form sulphur dioxide on burning. The enthalpy of combustion of rhombic sulphur is -296.6 kJ mol^{-1} and that of monoclinic sulphur is

$-297\,kJ\,mol^{-1}$. Suggest why these two values differ and deduce which of the allotropes is the more stable thermodynamically. Show the results of your deduction on an enthalpy diagram.

Calculate the enthalpy change for the reaction:

$$S(rhombic) \rightarrow S(monoclinic)$$

5 Data:

Substance	$H_2O(l)$	$CO_2(g)$	ethane, $C_2H_6(g)$	ethene, $C_2H_4(g)$
$\Delta H_f^{\ominus}/kJ\,mol^{-1}$	−285.5	−393	−83.6	+52.0

(a) Write equations for the complete combustion of:

(i) ethane;

(ii) ethene;

(iii) hydrogen.

(b) Calculate the enthalpy of combustion in each case.

(c) From the results obtained in (b), calculate the enthalpy change for the reaction:

$$C_2H_4(g) + H_2(g) \rightarrow C_2H_6(g)$$

6 Data: $4NH_3(g) + 5O_2(g) \rightarrow 4NO(g) + 6H_2O(l)$ $\Delta H = -1170\,kJ$

$4NH_3(g) + 3O_2(g) \rightarrow 2N_2(g) + 6H_2O(l)$ $\Delta H = -1530\,kJ$

$2H_2(g) + O_2(g) \rightarrow 2H_2O(l)$ $\Delta H = -576\,kJ$

(a) Calculate the enthalpy of formation of ammonia.

(b) Calculate the enthalpy of formation of nitrogen monoxide, $NO(g)$.

7 Data: $CH_4(g) + 2O_2(g) \rightarrow CO_2(g) + 2H_2O(l)$ $\Delta H = -890\,kJ\,mol^{-1}$.

$2CO(g) + O_2(g) \rightarrow 2CO_2(g)$ $\Delta H = -568\,kJ\,mol^{-1}$

$C(graphite) + O_2(g) \rightarrow CO_2(g)$ $\Delta H = -393\,kJ\,mol^{-1}$

$H_2(g) + \frac{1}{2}O_2(g) \rightarrow H_2O(l)$ $\Delta H = -285.5\,kJ\,mol^{-1}$

Use the data, as appropriate, to calculate:

(a) the enthalpy of formation of methane;

(b) the enthalpy of formation of carbon monoxide;

(c) the enthalpy change when 1 mole of methane burns in a limited supply of oxygen to produce carbon monoxide and water.

ENERGY – THE DRIVING FORCE OF LIFE

8 Data:

Substance	$B_2H_6(g)$	$B_2O_3(s)$	$C_6H_6(g)$	$CO_2(g)$	$H_2O(g)$
$\Delta H_f^\ominus$/kJ mol^{-1}	+31.4	−1270	+83.9	−393	−242

Gaseous diborane, B_2H_6, and gaseous benzene, C_6H_6, combust in oxygen as follows:

$$B_2H_6(g) + 3O_2(g) \rightarrow B_2O_3(s) + 3H_2O(g)$$

$$C_6H_6(g) + 7.5O_2(g) \rightarrow 6CO_2(g) + 3H_2O(g)$$

Calculate which will produce the greater amount of heat, 50 kg of $B_2H_6(g)$ or 100 kg of $C_6H_6(g)$.

9 Calculate the average bond enthalpy of a C–Cl bond given the following data:

$$C(graphite) \rightarrow C(g) \quad \Delta H = +715\,kJ\,mol^{-1}$$

$$Cl_2(g) \rightarrow 2Cl(g) \quad \Delta H = +242.2\,kJ\,mol^{-1}$$

$$C(graphite) + 2Cl_2(g) \rightarrow CCl_4(l) \quad \Delta H = -135.5\,kJ\,mol^{-1}$$

10 Data:

Bond	C–H	C–Br	Br–Br	H–Br
Average bond enthalpy/kJ mol^{-1}	413	209	193	366

Use the data to calculate the enthalpy change for the reaction:

$$CH_4(g) + Br_2(g) \rightarrow CH_3Br(g) + HBr(g)$$

11 The average bond enthalpies of hydrogen gas and nitrogen gas are 436 and 945 kJ mol^{-1}, respectively. Comment on the relative values of these average bond enthalpies and attempt to explain the difference.

The enthalpy of formation of ammonia gas is −46 kJ mol^{-1}.

Use the data to calculate the average bond enthalpy of the N–H bond.

12 The enthalpies of neutralisation of hydrofluoric acid (HF) and hydrocyanic acid (HCN) by 1 mol of sodium hydroxide are −68.6 kJ mol^{-1} and −11.7 kJ mol^{-1}, but both nitric and hydrochloric acid have enthalpies of neutralisation of −57.2 kJ mol^{-1}. Comment on this. (You will need to consult textbooks to find the properties of hydrofluoric and hydrocyanic acids).

2

Organic chemistry – general principles

What is organic chemistry?

The element carbon is able to form a vast number of compounds when in combination with hydrogen alone, and still more when other elements, such as oxygen, nitrogen and halogens, are introduced into the molecules.

Carbon has the ability to **catenate** (make chains), that is, the ability to form covalent bonds with itself which are particularly stable, as shown by the bond energies in Table 2.1. Hydrocarbon chains form the basis of most organic molecules. Stable molecules containing long chains of carbon atoms can form, as can certain ring structures. Carbon is unique in the extent to which catenation can occur. Organic compounds are the main constituents of all animal and plant life but they are **thermodynamically unstable** in the presence of oxygen, and combustion reactions are exothermic, for example:

$$CH_4(l) + 2O_2(g) \rightarrow CO_2(g) + 2H_2O(g) \quad \Delta H^\ominus = -890 \text{ kJ mol}^{-1}$$

DEFINITION

Organic chemistry is the study of the chemistry of the compounds of carbon.

Table 2.1 *Some bond energies compared to the C–C bond*

Bond	Bond energy/kJ mol^{-1}
C–C	347
N–N	158
O–O	144
Si–Si	226

Life on Earth therefore exists despite a very 'hostile' environment and would appear to be in imminent danger of spontaneous combustion. Fortunately, the activation energies of the reactions with oxygen are very high and so organic compounds are **kinetically stable** (see Chapter 6) at the sort of temperatures encountered on Earth.

A few compounds of carbon such as the oxides, carbonates and chlorides are not usually studied as organic chemistry and will be considered in the appropriate sections elsewhere.

Homologous series

The ability of carbon to form chains, continuous or branched, of almost any length means that the number of organic compounds is limitless. The study of such a vast number of compounds would be well nigh impossible were it not for the fact that the compounds can be arranged into a relatively small number of groups or families known as **homologous series**. These have certain features in common:
- there is a general formula for the series
- the molecular formulae of successive members differ by CH_2
- all members have similar chemical properties
- there is a gradual variation (a gradation) in physical properties.

ORGANIC CHEMISTRY – GENERAL PRINCIPLES

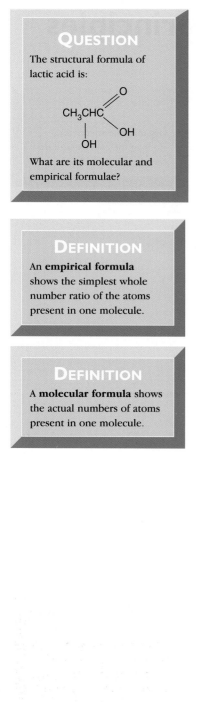

QUESTION

The structural formula of lactic acid is:

What are its molecular and empirical formulae?

DEFINITION

An **empirical formula** shows the simplest whole number ratio of the atoms present in one molecule.

DEFINITION

A **molecular formula** shows the actual numbers of atoms present in one molecule.

The reactions of one simple member of a series can be considered to be representative of the reactions of all members of the series. The amount of learning necessary is therefore considerably reduced. A variety of homologous series will be studied in this and later books.

Types of formulae

You will meet three main types of formulae in the study of organic compounds.

- **Empirical formula** – the formula which shows the simplest whole number ratio of the atoms of different elements present in one molecule.
- **Molecular formula** – the formula which shows the actual numbers of atoms of each element in one molecule.
- **Structural formula** – the formula which shows how the various atoms are bonded to one other within the molecule.

The **empirical formula** is only of use because it can be calculated from a knowledge of the percentage composition by mass of the compound, which in turn can be found from an analysis of the compound. Its importance is diminishing as spectroscopic methods (see Unit 5) give the molecular formula directly.

For example, for a compound which contains 82.75% carbon and 17.25% hydrogen by mass, the empirical formula can be calculated to be C_2H_5. The method for doing this was shown in Unit 1.

The molecular formula is a whole number multiple of the empirical formula. Thus it must be $(C_2H_5)_n$, where n is a positive integer, and the relative molecular mass of the molecule must be $[(2 \times 12) + (5 \times 1)] \times n = 29n$. The actual relative molecular mass of the compound must then be found experimentally. This can be done in a number of ways, including mass spectrometry using the mass of the molecular ion. In the example above the relative molecular mass is found to be 58. Hence $29n = 58$ and the value of n is 2. The molecular formula is therefore C_4H_{10}.

The above method dates back to the days when the molecular mass was very difficult to determine and values were often uncertain. Here, a value of 56, say, would still give $n = 2$ (it has to be a whole number). If a reasonably accurate value of the molecular mass is known, it is often easier to ignore the empirical formula and to calculate the molecular formula as follows:

$$\text{Number of C atoms} = \frac{82.75}{12} \times \frac{58}{100} = 4$$

$$\text{Number of H atoms} = \frac{17.25}{1} \times \frac{58}{100} = 10$$

$$\text{Molecular formula} = C_4H_{10}$$

Determination of the structural formula requires more information either from spectrometry or from a knowledge of the reactions of the compound.

Bonding in organic compounds

The bonding in organic compounds is almost always covalent, that is, one pair of shared electrons between carbon atoms. Double and even triple covalent bonds (involving two or three pairs of shared electrons) are also commonly found. 'Dot and cross' representations of the bonding in some simple molecules are:

methane   ethane

The covalent bond is more usually represented by a single line, thus:

It must be said at the outset that the molecules are not flat and the bond angles are not 90°. However, such representation shows which atoms are joined together.

Carbon atoms always show a covalency of four in organic molecules. As a result one possible structural formula for the molecular formula C_4H_{10} is:

$$H - \underset{\underset{H}{|}}{\overset{\overset{H}{|}}{C}} - \underset{\underset{H}{|}}{\overset{\overset{H}{|}}{C}} - \underset{\underset{H}{|}}{\overset{\overset{H}{|}}{C}} - \underset{\underset{H}{|}}{\overset{\overset{H}{|}}{C}} - H$$

The method of representation is often described as 'showing all covalent bonds'. A more condensed way of writing the formula that still shows the groupings around each carbon atom, and hence its structure, is $CH_3CH_2CH_2CH_3$.

Structural isomerism

Structural isomerism occurs when two or more different *structural* formulae can be written for the same *molecular* formula.

For the molecule C_4H_{10} another perfectly plausible structure could be:

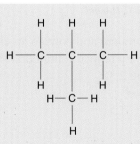

The condensed version could be written as $CH_3CH(CH_3)_2$. These two structures are said to be **structural isomers**. There are several ways in which

DEFINITION

Structural isomerism occurs when two or more different structural formulae can be written for the same molecular formula.

this structural isomerism can occur and these will be dealt with in the relevant places in the text. Each structure must have a different name in order to identify it. With such a vast number of possible structures, some systematic way of doing this is required and this will be introduced gradually as it applies to the different homologous series.

An orbital view of bonding

A more advanced theory of covalent bonding than that given earlier is based on the atomic orbital approach. A covalent bond is considered to be formed by the overlap of two atomic orbitals, each containing a single electron which must be of opposite spin. This results in the formation of a molecular orbital containing the shared pair of electrons which is a single covalent bond. The greater the degree of overlap between the two atomic orbitals, the stronger the covalent bond which is formed.

The electronic structure of a carbon atom in its ground state is $1s^2 2s^2 2p^2$ and there are only two unpaired electrons available for bonding. If one electron is promoted from the 2s orbital to the 2p orbital, four unpaired electrons become available (Figure 2.1). The energy required to do this is more than compensated for by the energy released in the formation of four bonds instead of two.

These electrons are not all equivalent since they are in different types of orbital. Thus when they each overlap with the 1s atomic orbital of a hydrogen atom, as in methane, three of the bonds would be of different length from the other one.

The formation of single covalent bonds

Consider the simple molecule of methane CH_4 which has four single bonds symmetrically arranged. The bonding can be understood if the excited carbon atom is considered to have its four orbitals rearranged into four new orbitals which are all equivalent. This process is referred to as **hybridisation** and the resulting orbitals as **sp^3 hybrid orbitals**. The shape of the hybrid orbitals is similar to a p atomic orbital except that one lobe is bigger than the other. It is the larger lobe which overlaps linearly with the 1s atomic orbital of a hydrogen atom forming a so-called σ (or sigma) bond as shown in Figure 2.2.

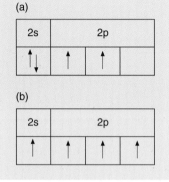

(a)

2s	2p		
↑↓	↑	↑	

(b)

2s	2p		
↑	↑	↑	↑

Fig. 2.1 Electronic arrangements in the second energy level for carbon atoms: (a) in the ground state and (b) in the excited state.

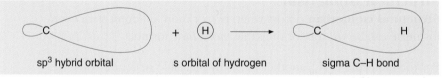

| sp³ hybrid orbital | s orbital of hydrogen | sigma C–H bond |

Fig. 2.2 The formation of a sigma C–H bond.

Thus all C–H bonds are equivalent and contain an electron pair. These electron pairs repel each other to give a regular tetrahedral shape to the methane molecule, as shown in Figure 2.3.

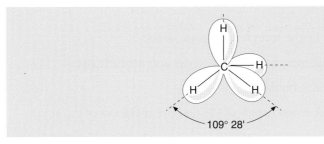

Fig. 2.3 The tetrahedral shape of the methane molecule.

This type of hybridisation occurs in all compounds where carbon atoms form four single covalent bonds and the arrangement of the bonds around each carbon atom is tetrahedral.

The formation of double covalent bonds

The carbon atoms which take part in double bond formation, as in ethene C_2H_4, undergo a different type of hybridisation. In this case the 2s and *two* of the 2p orbitals of the excited carbon atom are hybridised to form *three* new hybrid orbitals known as **sp^2 hybrid** orbitals. These are the same shape as the sp^3 hybrid orbitals but they lie in one plane at angles of 120°. This leaves one p orbital on the carbon atom, which lies at right angles to the plane of the sp^2 hybrid orbitals and which contains a single unpaired electron. When two such atoms combine to form the molecule of ethene, linear overlap between an sp^2 hybrid orbital from each carbon atom leads to the formation of a sigma bond between the two carbon atoms.

The other two sp^2 hybrid orbitals on each carbon atom can then undergo linear overlap with the 1s orbitals of four hydrogen atoms giving rise to four sigma bonds between the carbon and hydrogen atoms. The basic C_2H_4 molecule is thus formed and all the sigma bonds are in the same plane and at angles of 120° to each other, as shown in Figure 2.4.

The remaining p orbitals are at 90° to this plane and the positions are such that they are able to undergo lateral overlap leading to a different kind of covalent bond known as a π (or pi) bond (see Figure 2.5). This comprises two 'sausage-shaped' regions of negative charge, one above and one below the plane of the sigma bonds. This therefore explains the formation of the second bond between the carbon atoms which is completely different from the first. It is the formation of this π-bond which holds the σ-bonds in the same plane.

Types of reagent

Most types of reagent create new covalent bonds. There are two main types of reagent:
- **nucleophiles** – species which seek out positive centres and must have a lone pair of electrons which they can donate to form a new covalent bond.
- **electrophiles** – species which seek out negative centres and must be capable of accepting a lone pair of electrons to form a new covalent bond.

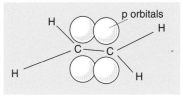

Fig. 2.4 Two carbon atoms linked by overlap of sp^2 hybrid orbitals (shown as lines). The p orbitals are at right angles to the plane of the hybrid orbitals.

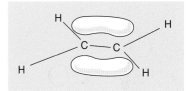

Fig. 2.5 The pi bond (π-bond) formed by lateral overlap of the non-hybridised p atomic orbitals of carbon.

DEFINITION

Nucleophiles are species which seek out positive centres and must have a lone pair of electrons which they can donate to form a new covalent bond.

DEFINITION

Electrophiles are species which seek out negative centres and must be capable of accepting a lone pair of electrons to form a new covalent bond.

ORGANIC CHEMISTRY – GENERAL PRINCIPLES

Questions

1 *(a)* What is meant by the term *homologous series*?

  (b) Look up five different homologous series and for each give:

   (i) the general formula;

   (ii) the formulae and names of the first five members of the series.

2 *(a)* What is meant by the term *structural isomerism*?

  (b) Deduce the structural formulae of all isomers of C_5H_{12}.

3 Describe in orbital terms how carbon atoms can form:

  (a) single covalent bonds with each other;

  (b) double covalent bonds with each other.

4 *(a)* What is meant by the terms *empirical formula* and *molecular formula*?

  (b) A compound contains 52.2% carbon, 13% hydrogen and 34.8% oxygen by weight and has a relative molecular mass of 46. Calculate its empirical and molecular formulae.

   Attempt to write two different structures for this molecule.

5 A compound contains 12.8% C, 2.1% H and 85.1% Br. What is its empirical formula? Its molecular mass is in the range 150–250. What is its molecular formula? Write two possible structures for the compound. Why is knowledge of the molecular mass of the compound not necessary to answer the question?

3

Alkanes and alkenes

Aliphatic and aromatic compounds

In the nineteenth century, chemists tried to classify or make sense of an ever-increasing number of known organic compounds. They found that some compounds were waxy or fatty and some had prominent (and often pleasant) smells. Chemists called these two groups of compounds **aliphatic** (fatty) and **aromatic**. This system of classification, essentially based on physical properties, was unsound, but the two terms were retained with different meaning. Until you know more organic chemistry it is not possible to understand the modern meaning of the terms. It may help you to know that aromatic compounds always contain rings (often of six carbon atoms). They appear from their molecular formulae to have many double bonds, e.g. benzene has the formula C_6H_6. Six carbon atoms in a chain, connected to one another and to hydrogen by single bonds, would give a compound with the formula C_6H_{14}. Aromatic compounds do not show the properties expected of compounds with double bonds, however. The majority of aliphatic compounds, on the other hand, do not contain rings; if they contain double bonds, they show the expected reactions.

Alkanes

The compounds known as **alkanes** form what is generally regarded as the simplest homologous series. They are the source of very many organic chemicals and are of great commercial and economic importance. They also form the basis of the systematic nomenclature for all other aliphatic organic substances.

General formula

C_nH_{2n+2}

Members and nomenclature

The formulae of alkanes are simply obtained by inserting ascending integral values for n, beginning with 1. The first ten members are shown in Table 3.1.

The names all follow the general name for the series and end with the suffix **-ane**. The prefix **alk-** changes to indicate the number of carbon atoms in the molecule. From pentane onwards the prefix is derived from the Greek, but the first four retain the names originally given to them. In the systematic nomenclature, **meth-**, **eth-**, **prop-** and **but-** will always refer to chains of one, two, three and four carbon atoms, respectively.

The molecular formula shows an increase of CH_2 from one member to the next, and this is the **homologous increment**. As a result, the relative molecular masses increase by steps of 14 as the series is ascended.

Table 3.1 *The first ten alkanes*

n	Formula	Name
1	CH_4	methane
2	C_2H_6	ethane
3	C_3H_8	propane
4	C_4H_{10}	butane
5	C_5H_{12}	pentane
6	C_6H_{14}	hexane
7	C_7H_{16}	heptane
8	C_8H_{18}	octane
9	C_9H_{20}	nonane
10	$C_{10}H_{22}$	decane

Alkyl groups will be encountered when dealing with structural isomerism. The general formula for these is $-C_nH_{2n+1}$. They are not capable of independent existence but they do occur within other molecules. They are named in the same way as the corresponding alkane except that the name ends in **-yl** instead of -ane. The two most common ones are:
- $-CH_3$, the **methyl** group
- $-C_2H_5$ or $-CH_2CH_3$, the **ethyl** group.

Physical properties

The physical properties of the alkanes show a gradual variation as the homologous series is ascended. Table 3.2 shows the boiling temperatures and melting temperatures for the first eight members. A graph of boiling temperatures against the number of carbon atoms in the molecule shows a smooth increase. This is not true for a similar graph of melting temperatures against number of carbon atoms.

Table 3.2 *Boiling and melting temperatures of the first eight alkanes*

Name	Boiling temperature/°C	Melting temperature/°C
methane	−162	−182
ethane	−89	−183
propane	−42	−188
butane	−0.5	−138
pentane	36	−130
hexane	69	−95
heptane	98	−91
octane	126	−57

Fig. 3.1 Alkanes can occur naturally. North Sea natural gas is largely methane.

It is obvious from Table 3.2 that the alkanes gradually change their physical state at room temperature as relative molecular mass increases. Thus C_1 to C_4 are colourless gases, C_5 to C_{15} are colourless liquids and above C_{15} they are white waxy solids. They are very common materials in everyday life, for example, propane gas and butane gas are used as mobile sources of heat and light for camping, etc. They are actually sold in the liquid state under pressure in cylinders. Methane is piped into almost every home, directly from the North Sea, for cooking and heating purposes. The liquid alkanes are used in petrol for cars, paraffin, etc.

The majority of the uses outlined in the previous paragraph require the combustion of the alkane in order to release the energy from the molecules. It is important to realise, however, that alkanes are the main source of many organic materials, including many plastics, synthetic fibres, detergents, etc. Some inorganic materials are also manufactured from natural alkanes. Hydrogen (from processed alkanes) is used to make ammonia. Sulphur (as H_2S) and iodine occur in the impurities in gas and oil.

Fig. 3.2 Typical products containing alkanes.

Sources of alkanes

The main source of alkanes is from underground deposits of crude oil or natural gas, both of which require drilling in order to extract the raw materials. Sources are currently plentiful, but are thought to be finite.

QUESTION

Draw a graph to illustrate the data in Table 3.2. Use it to predict the boiling and melting temperatures of nonane and decane, the next members of the series.

Natural gas requires little purification before being used, but crude oil is a very complex mixture of alkanes and other hydrocarbons – which require separation, usually into groups of compounds which have similar boiling temperatures. This is achieved initially by fractional distillation of the crude oil.

Bonding in alkanes

The carbon atoms in alkanes are sp^3 hybridised and are joined to one another, and to the hydrogen atoms, by sigma bonds. Since each sigma bond contains a pair of electrons, the mutual repulsion results in a tetrahedral arrangement. Structural formulae often show the bonds in such a way that the molecule appears to be planar and the molecules are often referred to as 'straight chain' molecules, for example:

Where it is necessary or desirable to show the three-dimensional nature of the bonding, a dashed line (- - -) will be used to indicate a bond going *into* (or being behind) the page and a wedge-shape (◁) to indicate a bond *coming out* of (in front of) the page. Ordinary lines will indicate bonds *within* the plane of the page. Thus the two structures above could be shown as follows:

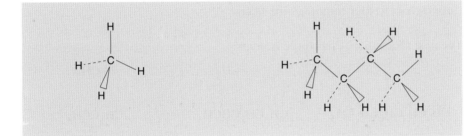

Structural isomerism

In alkanes, the only way in which different structures can be obtained is by rearranging the carbon chain. The structure in which all the carbon atoms are joined together in a continuous chain is known as the 'straight chain' isomer (although the molecule is not actually linear as just mentioned above). Since the structural formula does not normally represent the exact shape of the molecule, bending the chain does not actually change the 'length' of the carbon chain.

Other isomers are formed when some carbon atoms are not part of the main chain but form 'side chains' of different lengths. These are referred to as **branched chain** molecules. The side chains can never be of greater length than the parts of the main chain on either side of the carbon to which they are joined. To illustrate this point, if you had a pentane molecule $CH_3CH_2CH_2CH_2CH_3$, for example, and you attached a group to the central carbon atom, there would be a two-carbon (ethyl) group on either side. You cannot attach a bigger group than ethyl in naming this example. A methyl group would give 3-methylpentane, an ethyl group would give 3-ethylpentane but on adding a propyl group, $CH_3CH_2CH_2-$, it must not be called 3-propylpentane, because propyl contains more carbon atoms than either part of the main chain attached to carbon-3. Instead it should be called 3-ethylhexane. Draw out the pentane molecule, alter it by adding a propyl group, and then rearrange it to check this.

The first alkane to show isomerism is C_4H_{10}, which can have two structures:

$$CH_3CH_2CH_2CH_3 \quad \text{and} \quad \overset{\overset{\textstyle CH_3}{\textstyle |}}{CH_3CHCH_3} \qquad \text{(or } CH_3CH(CH_3)_2)$$

$$\text{butane} \qquad\qquad \text{2-methylpropane}$$

The **systematic nomenclature** for organic compounds uses a name based on the longest continuous carbon chain. Alkyl groups are regarded as substituents and their names come before the name of the longest carbon chain, and they are preceded by a number to indicate the carbon atom at which the substitution has occurred. The number does *not* refer to the number of alkyl groups being substituted.

Thus the name 2-methylpropane indicates that a methyl group has been substituted on the second carbon atom of a three-carbon chain. The prefix 2- is not strictly necessary in this case since the second carbon atom is the only atom where substitution by a methyl group could occur without increasing the carbon chain to four carbon atoms again.

The next alkane C_5H_{12} has three isomers:

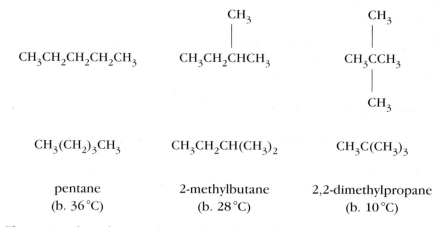

$$CH_3CH_2CH_2CH_2CH_3 \qquad \overset{\overset{\textstyle CH_3}{\textstyle |}}{CH_3CH_2CHCH_3} \qquad \overset{\overset{\textstyle CH_3}{\textstyle |}}{\underset{\underset{\textstyle CH_3}{\textstyle |}}{CH_3CCH_3}}$$

$$CH_3(CH_2)_3CH_3 \qquad CH_3CH_2CH(CH_3)_2 \qquad CH_3C(CH_3)_3$$

| pentane | 2-methylbutane | 2,2-dimethylpropane |
| (b. 36°C) | (b. 28°C) | (b. 10°C) |

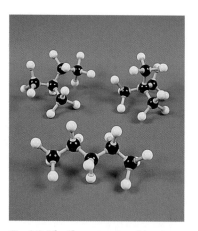

Fig. 3.3 The three structural isomers of C_5H_{12}.

The main carbon chain can be numbered from either end, and this should be done so as to produce the lowest number for the substituents (2-methylbutane rather than 3-methylbutane in the example above). Where there is more than one substituent, each one must be given a number. Any name used must be unambiguous. It would be good practice to make models of these different structures to see how the overall shape varies.

All isomers of alkanes have similar chemical properties but differ in physical properties such as boiling temperature, and it is this that allows individual isomers to be isolated (i.e. alkanes isolated in the pure state). The boiling temperature is determined by the strength of the intermolecular forces of attraction in the liquid state. The only intermolecular forces present in alkanes are van der Waals' forces and the strength of these depends on the proximity with which the molecules can approach one another. The intermolecular interaction is greatest in straight chain molecules and decreases as the degree of branching increases. Hence boiling temperatures decrease as degree of branching increases, as shown in the examples above.

Cyclic alkanes

A number of alkane molecules can be formed where the ends of the carbon chain have been linked up to form a 'ring' or 'cyclic' structure, for example:

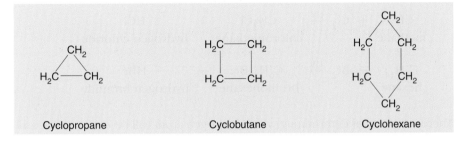

| Cyclopropane | Cyclobutane | Cyclohexane |

The bond angles for the smaller rings are very different from the 109° 28' of the sp³ hybrid orbitals, which indicates poor overlap of the orbitals and considerable strain in the ring. They behave as normal alkanes as far as the C–H bond is concerned, but would be much more reactive towards reagents which break the C–C bond. They do not have the general formula of simple alkanes.

Reactions of alkanes

Combustion

All alkanes, and indeed all hydrocarbons, will burn in air or oxygen with the release of heat energy and many are used for this purpose. The products of the complete combustion in excess oxygen are always carbon dioxide and water. As a result, equations can be simply constructed; for example, the equation for combustion of methane is:

$$CH_4(g) + 2O_2(g) \rightarrow CO_2(g) + 2H_2O(g) \quad \Delta H = -882 \, kJ \, mol^{-1}$$

The same kind of reaction is used to heat homes, drive motor cars, power aeroplanes, and for many other purposes.

If the alkane and oxygen are mixed first, an explosion will result on ignition. Great care needs to be taken to avoid this in coal mines where methane is released naturally and can build up to explosive proportions. In car engines however just such a small explosion is carried out in order to drive the engine.

Halogenation

Halogenation is the introduction of a halogen atom into the alkane molecule. It occurs by a **substitution reaction**, which means that a halogen atom replaces one or more of the hydrogen atoms in the alkane. Hydrogen halides are always produced in these reactions (as 'steamy' acidic fumes) together with a product called a **haloalkane**.

> **QUESTION**
>
> Write a general equation for the combustion of any alkane (C_nH_{2n+2}). Which gaseous alkane will require five times its volume of oxygen to burn completely?

> **DEFINITION**
>
> A **substitution reaction** is one in which an atom or group of atoms in one molecule is replaced by another atom or group of atoms.

All halogens react with all alkanes, but the reaction is quicker with chlorine than with bromine which is in turn quicker than with iodine. Chlorine may react explosively but iodine results in an equilibrium. Also, the rate of reaction decreases as the relative molecular mass of the alkane increases. The presence of sunlight, ultraviolet (UV) light or some other form of energy is essential for these reactions to proceed at a reasonable rate. This can easily be demonstrated experimentally using hexane and bromine and placing the same mixture in two test tubes, one of which is wrapped in black paper. Exposure to sunlight or a strong artificial light leads to a rapid decolorisation of the bromine in the uncovered test tube while the colour remains for much longer in the darkened one. Examples of such reactions are:

$$CH_4 \ + \ Cl_2 \ \overset{UV}{\rightarrow} \ CH_3Cl \ + \ HCl$$

methane   chloromethane   hydrogen chloride

$$C_6H_{14} \ + \ Br_2 \ \overset{UV}{\rightarrow} \ C_6H_{13}Br \ + \ HBr$$

hexane   bromohexane   hydrogen bromide

There is no way of determining which hydrogen atom will be replaced since all the C–H bonds are equivalent. It is not possible therefore to make a specific haloalkane by this method since it is impossible to stop further substitutions taking place.

These reactions are **free-radical** in nature. They involve **homolytic bond fission** which is, in effect, 'undoing' a covalent bond.

$$X : Y \rightarrow X \cdot + \cdot Y$$

You already know of the high values of bond energies and will realise that, in practice, this process is not very easy. It is usually initiated by the action of light or at very high temperatures. Given the bond energies $E(C–H) = 413$ kJ mol^{-1} and $E(Cl–Cl) = 242$ kJ mol^{-1}, it is clear that the Cl–Cl bond will be the first to be broken. Once this has happened, the active free radicals then begin a process of breaking other bonds

$$C : H + Cl \cdot \rightarrow C \cdot + H : Cl$$

But since there is an odd electron, there will always be a free-radical product. This chain reaction then continues until two of the free radicals can rejoin. Once started, such reactions can become very vigorous.

Alkenes

This is another homologous series containing carbon and hydrogen only.

General formula

$$C_nH_{2n}$$

Each molecule contains two hydrogen atoms less than the corresponding alkane.

> ### DEFINITION
> **Free radicals** are species which have a single unpaired electron.

> ### QUESTION
> Why is it possible to identify a simple alkane but not a simple alkene from its empirical formula?

Members and nomenclature

The names of the alkenes follow exactly from the general name for the series. Thus, compared to alkanes, the suffix simply changes to **-ene** whilst the prefix remains the same, indicating the number of carbon atoms. There is no alkene corresponding to $n = 1$, so the first two members of the series are:

ethene C_2H_4 or $H_2C=CH_2$
propene C_3H_6 or $CH_3CH=CH_2$

All alkene molecules contain one carbon–carbon double bond at some point in the carbon chain and as a result are said to be **unsaturated**.

Structural isomerism

Structural isomerism can occur by moving the double bond to different positions in the carbon chain. The position of the double bond is then indicated by a number inserted between the prefix and the -ene, for example C_4H_8 could be

$CH_3CH_2CH=CH_2$ or $CH_3CH=CHCH_3$
but-1-ene   but-2-ene

The number used is the smallest one possible, counting from each end in turn, and using the lower number of the carbon atom in the pair joined by the double bond.

Branching of the carbon chain is still possible as the length of the carbon chain increases. In these cases, the names are still based on the longest continuous carbon chain present, as shown by *some* of the isomers of C_5H_{10} below:

$CH_3CH_2CH_2CH=CH_2$ $(CH_3)_2CHCH=CH_2$ $(CH_3)_2C=CHCH_3$
pent-1-ene   3-methylbut-1-ene   2-methylbut-2-ene

Note that the position of the double bond takes precedence in numbering the carbon atoms of the longest carbon chain. As a result, the isomer 3-methylbut-1-ene is not called 2-methylbut-3-ene.

Such structural isomers differ from each other only in physical properties; their chemical reactions are effectively the same. Apart from this structural isomerism, a new kind of isomerism (geometric isomerism) occurs in alkenes, which is discussed later.

Bonding in alkenes

The two carbon atoms of the double bond are sp^2 hybridised and form a double covalent bond with each other, one bond of which is a sigma bond and the other a pi bond. The arrangement of the bonds around these two carbon atoms is planar with bond angles of 120°.

The pi bond between the carbon atoms is therefore completely different from the sigma bond and consists of two regions of negative charge, one above and one below the plane of the carbon and hydrogen atoms (see Fig 2.5, p.27).

Experimental evidence which indicates that the two bonds in a carbon–carbon double bond are not the same includes the following.

- The bond energy of C=C ($612\,\text{kJ mol}^{-1}$) is greater than C–C ($348\,\text{kJ mol}^{-1}$) but not twice as big. Hence the pi bond is weaker than the sigma bond.
- The greater strength of the C=C bond is supported by the shorter bond length (hence greater overlap) which is 0.134 nm as opposed to 0.154 nm for C–C.
- The existence of geometric isomers, as explained below.

Carbon atoms in the alkene molecule other than the two directly engaged in the formation of the double bond will probably be sp³ hybridised and the arrangement of the bonds around these carbon atoms will be tetrahedral as usual.

Geometric isomerism

Geometric isomerism is a different kind of isomerism which exists as a direct consequence of the double bond. A single sigma bond between carbon atoms will allow rotation around the axis of the bond without any reduction in the degree of overlap. As a result, there is free rotation about this bond since no covalent bonds need to be broken. With a double bond however, such rotation would lead to a decrease in the degree of overlap of the pi bond and consequently cannot be achieved without the supply of a suitable quantity of energy. There is restricted rotation about the carbon–carbon double bond.

The isomerism depends on the arrangement of the groups or atoms around the double bond, for example, but-2-ene can have two structures:

cis–but–2–ene　　　　trans–but–2–ene

It is difficult to describe, in words alone, the difference between *cis-* and *trans-* isomers. In essence, two substituents, one on each of the two carbon atoms whose rotation is restricted, are situated as near as possible to each other (*cis-*) or as far away as possible (*trans-*).

The general term for isomerism relying on the orientation (arrangement in space) of groups is **stereoisomerism** and geometric isomerism is one form of this. Such isomers have only very slight differences in physical properties such as boiling temperatures.

This kind of isomerism is to be found in any molecule of the type:

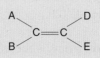

where A is not the same atom or group as B and D is not the same as E. Some further examples are:

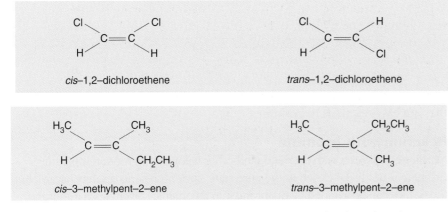

cis–1,2–dichloroethene   trans–1,2–dichloroethene

cis–3–methylpent–2–ene   trans–3–methylpent–2–ene

Geometrical isomerism can occur in saturated compounds where bond rotation is restricted by the formation of a ring.

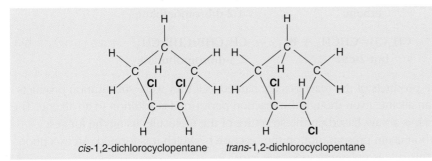

cis-1,2-dichlorocyclopentane   trans-1,2-dichlorocyclopentane

You do not need to recall or know this for the Edexcel exam but you might be expected to deduce it.

Reactions of alkenes

As shown earlier, the bonding in alkenes is quite different from that in alkanes and it is therefore not surprising that their reactions are also quite different. The pi (π) bond in alkenes is much more available to attack and is also a weaker bond than the sigma (σ) bonds in alkanes. Hence the alkenes will react with a greater variety of reagents and will react much more readily than alkanes.

Addition reactions

Addition reactions are the most typical of the alkene reactions, although they are not the only kind of reaction which alkenes undergo. An addition reaction is one in which two molecules react together to form a single product.

In the addition reactions of alkenes, it is always the pi bond that breaks in order to release electrons to form bonds with the reactant molecule. Because the reaction occurs at this point of high electron density, it is further qualified by the use of the term **electrophilic addition**. The attacking (electron seeking) species, e.g. bromine (see below) is called an **electrophile**.

The sigma bonds remain intact so that the two carbon atoms of the double bond always remain bonded together and a saturated molecule is formed from an unsaturated alkene. The equations can therefore always be written as follows:

> **QUESTION**
>
> Would you expect a species which attacked alkenes to be an electrophile or a nucleophile? Explain your choice.

> **DEFINITION**
>
> An **addition reaction** is a reaction in which two molecules react together to form a single product.

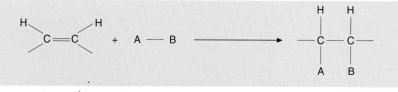

These reactions often occur at room temperature, which is an indication that they are easy reactions to perform.

Reaction with bromine

All alkenes will react with bromine (and other halogens) at room temperature. Bromine is not usually used in the pure state since it presents a safety hazard, but the hazard can be avoided by dissolving it in an organic solvent such as hexane. Typical reactions are as follows:

$$H_2C=CH_2 + Br_2 \rightarrow CH_2BrCH_2Br$$
ethene $\qquad\qquad$ 1,2-dibromoethane

$$CH_3CH=CHCH_3 + Br_2 \rightarrow CH_3CHBrCHBrCH_3$$
but-2-ene $\qquad\qquad$ 2,3-dibromobutane

The products of the reactions are named as if they were substitution products of an alkane, even though the reaction performed is addition to an alkene. The name is always based on the structure of the molecule, using the longest carbon chain present, not on the reaction by which it is formed. An exception to this rule is polymerisation, see later.

Reaction with potassium manganate(VII)

Another reaction which occurs easily is the oxidation of alkenes by alkaline potassium manganate(VII).

$$H_2C=CH_2 + 2OH^- - 2\ e^- \rightarrow HOCH_2CH_2OH$$
ethane-1,2-diol

The purple colour of the manganate(VII) rapidly goes green or brown as it is reduced to manganate(VI) or manganese(IV) oxide. The product here is widely used as 'glycol antifreeze' but it is not made using expensive potassium manganate(VII) as the oxidising agent. If you perform this reaction as a test for unsaturation on a liquid alkene, e.g. turpentine, you must remember that hydrocarbons do not mix with water and thus require shaking thoroughly.

Tests for unsaturation

If a reaction is to be used as a test it should, generally speaking, be capable of producing an observable result in a reasonable time and in a relatively simple piece of apparatus.

Reduction with hydrogen

Reduction with hydrogen – paradoxically known as an 'addition reaction' here – occurs only with the aid of a catalyst, and usually requires elevated temperatures and pressures. Common catalysts are (finely divided) nickel, platinum and palladium.

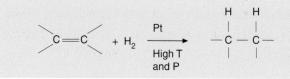

The reaction is particularly important commercially (where the cheaper nickel catalyst is used) to covert unsaturated oils (liquids) into low-melting saturated fats for use in margarine.

Reaction with hydrogen bromide

The reaction here is exactly as predicted from the general equation. Hence the reaction with ethene at room temperature is as follows:

$$H_2C=CH_2 + HBr \rightarrow CH_3CH_2Br$$
bromoethane
(a colourless liquid)

Unlike the previous examples, in which the *same* atom or group was added to each of the carbon atoms joined by the double bond, a possibility of isomerism can occur when adding a hydrogen halide. This happens when the alkene is unsymmetrical, i.e. the groups at each end of the double bond are not identical. Thus but-1-ene gives two products with hydrogen bromide

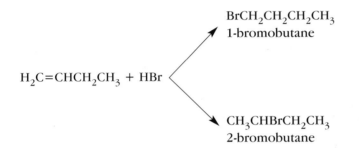

$$BrCH_2CH_2CH_2CH_3$$
1-bromobutane

$$H_2C=CHCH_2CH_3 + HBr$$

$$CH_3CHBrCH_2CH_3$$
2-bromobutane

In this case, 2-bromobutane predominates. It was discovered by Markownikov that in such reactions the hydrogen of the HBr was attached to the carbon atom carrying more hydrogen atoms. You will learn the reason for this in a later unit; until then Markownikov's rule will guide you.

Poly(ethene)

Poly(ethene) is a **polymer**, that is, a very large molecule and, since it is made from a single unit or **monomer** by a process of **addition**, it is known as an **addition polymer**.

The single monomer unit is ethene, molecules of which are made to join together (by a free-radical mechanism) to form very long molecules which are essentially alkanes. The basic reaction is usually represented as

$$n CH_2=CH_2 \rightarrow (-CH_2-CH_2-)_n$$

where $(-CH_2-CH_2-)_n$ is called the repeating unit.

Poly(ethene), more commonly known as 'polythene', is a plastic material which is in everyday use for many different purposes such as film packaging, electrical insulation, containers for household chemicals such as washing-up liquid, buckets, food boxes, washing-up bowls, etc. This is obviously a very important industrial application of alkenes.

In fact, there are different forms of poly(ethene), the two main ones being: low density poly(ethene), LDPE, and high-density poly(ethene), HDPE.

Fig. 3.4 Low-density polythene is widely used for packaging.

Fig. 3.5 High-density polythene has a structure of considerable strength.

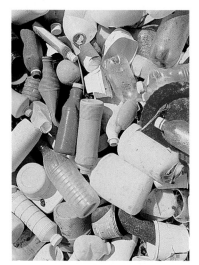

Fig. 3.6 Polythene is not biodegradable, and when burned gives off toxic fumes. This causes problems of disposal with long-term effects.

Fig. 3.7 Poly(propene) has high tensile strength and resists wear well – essential in fishing nets.

Low-density poly(ethene) is made by subjecting the ethene to a very high pressure (1000–3000 atm) at a moderate temperature in the range 420–570 K. The average polymer molecule contains between 4×10^3 and 40×10^3 carbon atoms, with many short carbon chain branches.

High density poly(ethene) is made by subjecting ethene to a lower temperature (310–360 K) and pressure (1–50 atm) in a suspension of titanium(III) or titanium(IV) chloride and an alkylaluminium compound such as triethyl aluminium (known as a Ziegler–Natta catalyst). This type of poly(ethene) has few branched chains and the molecules can pack more closely together. It is therefore more dense and melts at a higher temperature. It has a crystalline structure which gives it a rigid structure of considerable strength.

Poly(propene)

Polymers other than poly(ethene) are capable of wide structural variation, in addition to the branching characteristics of LDPE. Polymerisation of propene was not commercially successful until the use of catalysts of the Ziegler–Natta type produced **stereoregular** polymers.

The structural regularity is not limited to the presence of a methyl side chain on every **alternate** carbon atom

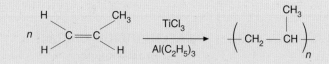

but the geometrical arrangement of the methyl groups in space is regular also; for example, this arrangement is described as isotactic:

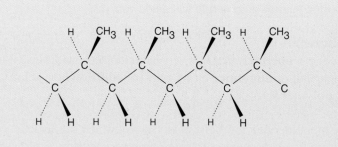

In reality, the methyl groups are arranged in a spiral which stiffens the structure and allows the long molecules to line up close to one another. This, in turn, increases hardness and wear-resistance and raises the softening temperature and strength of the material (compare HDPE). It is widely used to make ropes, sacking and carpets.

Poly(chloroethene) – PVC

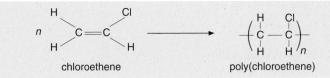

chloroethene → poly(chloroethene)

This is widely used for purposes as diverse as plastic handbags and rainwear, when it is softened by the addition of oils called plasticisers. For floor tiles it is hardened by the addition of mineral fillers. It revolutionised the manufacture of electric cable in the second half of the twentieth century, largely displacing rubber and cotton. Its chief disadvantage as an electric insulator is that in the event of an 'electric fire' it can melt causing short circuits, and at high temperatures, especially in the presence of air, it can give rise to some very toxic chlorine-containing compounds.

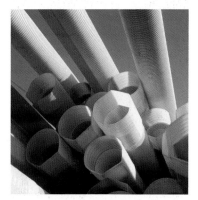

Fig. 3.8 Rigidity and resistance to wear make poly(chloroethene) an excellent choice for pipes.

Poly(fluoroethene) – PTFE or Teflon

tetrafluoroethene → poly(tetrafluoroethene)

The C–F bond is almost chemically unattackable: $E(C–F) = 480$ kJ mol^{-1}. The molecule is rather like a tightly twisted rope and the resulting solid is quite hard and very slippery. It is used where a lubricated surface is in contact with chemicals for seals, burette taps and bearings for stirrers. A thin ribbon is used by plumbers to seal the threads of screw-joints in piping and, of course, it is used as a surface coating for some non-stick ovenware.

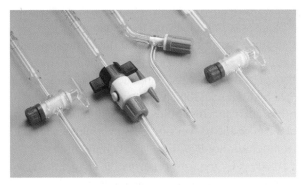

Fig. 3.9 Poly(tetrafluoroethene), or PTFE, resists chemical attack well. The taps on these burettes are made from it.

The problem of disposal

Polythene, along with many other polymers, provides us with many very useful materials. It does, however, present a considerable environmental problem when we come to dispose of it since it is not biodegradable, that is, it is not broken down by the action of microorganisms. As a result, it accumulates in vast quantities in rubbish tips and will never disappear. This is a problem which needs to be solved in the near future if we are not to leave an ever-increasing problem for future generations.

Liquid and gaseous fuels

Among other properties, an ideal fuel should:

1. be abundant
2. be easy and safe to store and move
3. be non-toxic
4. have a high calorific value
5. give rise to harmless combustion products.

In addition, it is helpful if the fuel is easily vaporised (not a problem with gaseous fuels) which, for liquids, causes an immediate conflict with easy storage.

Abundance: methane, butane and octane are limited resources obtained directly or indirectly from oil and gas wells. While all of them can be made from coal, this itself is a finite resource. Ethanol is continuously replaceable by fermentation of sugars of vegetable origin; however, if the Earth's energy needs were to be met by ethanol, it is debatable whether enough vegetable matter could be produced.

There is unlimited hydrogen, because it can be obtained by the electrolysis of water; what is more, when you burn it you convert it back to its source and when you regenerate it by electrolysis you recover the oxygen used to burn it as well. The snag to this vision of perfect happiness is that you require the fuel to produce energy but you will use all that energy (and more) to recover the fuel. Nuclear fission is one obvious, though currently unpalatable, answer to provide the energy for electrolysis; controlled nuclear fusion still has too many problems to solve.

Storage and transport: The transport of any liquid fuel has its obvious dangers but the storage and transport of gases is potentially far more dangerous. Hydrogen is particularly difficult since it cannot be liquefied under pressure at room temperature; the more you store in a given container the higher the pressure required and the stronger and heavier the container must be. A large industrial cylinder of 150 dm^3 of hydrogen at 2000 psi (pounds per square inch is still in use for gases in cylinders and car tyres) holds not much more than 1 kg of hydrogen (the mass of a bag of sugar), yet one person could not lift the cylinder. In a fire, raised to a temperature of 800–1000 °C, such a cylinder would generate a pressure of 10 000 psi – ten tons on every bit of area the size of a match box. If you have heard a car tyre explode at 30 psi or seen the devastating effect when a heavy goods vehicle has a 'blow-out' at 100 psi , scattering bits of tyre all over a motorway, you can imagine the effect of an exploding cylinder of hydrogen – a highly flammable gas, to make things worse. The most spectacular hydrogen fire ever was probably the Hindenberg (Figure 3.10); it is amazing that there were any survivors.

Scientists have given much thought to the safe storage of hydrogen. Storage and transport as a liquid (b. 20 K, –253 °C) is one difficult, but not impossible, solution. Liquid oxygen (b. 90 K, –183 °C) and liquid nitrogen

Fig. 3.10 The most spectacular hydrogen fire ever was probably the Hindenberg disaster.

(b. 77 K, −196 °C) are transported on our roads daily; however, their relatively light-weight but highly insulated containers can safely be allowed to 'leak away' any excess pressure. Absorption into metals such as palladium is a very safe and easy method. On gently heating, the hydrogen is released. The Earth's reserves of palladium, a very expensive metal, are minute, but perhaps an alloy can be found with similar properties.

Butane gas is easily liquefied at moderate pressures and normal ambient temperatures and the storage cylinders do not need to be so strong. In very cold weather, however, serious problems are caused when the liquid vaporises far too slowly. For camping and portable stoves, liquefied butane ('Calor' gas) is very convenient in the summer, but for outdoor storage for domestic use throughout the year, mixtures of propane and butane, or propane alone are used.

Toxicity: Hydrogen and methane can be breathed in small amounts without harm; butane, octane and most gaseous hydrocarbons have narcotic and hallucinogenic properties in small but dangerous amounts and are poisonous in larger quantities. Ethanol is readily imbibed in dilute solution by many of the population but it is poisonous in all but small amounts and, taken continuously, causes long-term damage to organs of the body.

Calorific value: Rightly, as a chemist, you should think of ΔH_c as significant in any discussion of fuel efficiency. However, ΔH_c is based on 1 mole of substance, and we usually buy liquid fuels by the litre (dm^3) or possibly by the kg. It is easy enough to convert the data, for example

$\Delta H_c(C_8H_{18})$ = –5510 kJ mol^{-1}; its density is 0.703 g cm^{-3}. Calculate the calorific value per gram and per cm^3.

> (1 mole) C_8H_{18} = 114 g
> 114 g of octane yields 5510 kJ on combustion
> 1 g yields 5510/114 = 48.3 kJ
> calorific value = 48 kJ g^{-1}
>
> 1 cm^3 of octane has a mass of 0.703 g
> 1 cm^3 yields 0.703 × 48.3 kJ = 34.0 kJ cm^{-3}
> calorific value = 34 kJ cm^{-3}

Such calculations for the common group of fuels give the results shown in Table 3.3; the densities of the liquids and gases , D_g and D_l , are given to only one significant figure because they are very temperature dependent.

Table 3.3 *Calorific values (cv) of some common fuels*

Fuel	M_r	$-\Delta H_c$/kJ mol^{-1}	cv/kJ g^{-1}	D_g/g cm^{-3}	D_l/g cm^{-3}	cv/kJ cm^{-3}
H_2(l)	2	242	120		0.07	8
H_2(g)	2	242	120	0.00009		0.001
CH_4(g)	16	890	56	0.0007		0.04
C_2H_5OH(l)	46	1370	30		0.8	24
C_4H_{10}(l)	58	2880	50		0.6	30
C_4H_{10}(g)	58	2880	50	0.003		0.15
C_8H_{18}(l)	114	5510	48		0.7	34

From Table 3.3 it is clear that, if the weight of the fuel matters (aircraft or rockets for example) then hydrogen appears to be the best fuel, but for vehicles with limited bulk-carrying power (petrol tanks), hydrocarbons of increasing molecular mass are superior. Uncompressed gases are clearly very inferior.

Combustion products: Ideally all these fuels give only water and carbon dioxide. Hydrocarbons can give rise to carbon monoxide which is very poisonous and to carbon (soot) if the fuel:air ratio is incorrect (rich) or if the air is not correctly mixed with the fuel. The higher the molecular mass of the fuel the less volatile it is. The more it tends to stay in airless droplets, the more this becomes a problem.

In internal combustion engines, further problems arise. It is claimed that even the output of some hydrogen-burning engines has unacceptable levels of hydrogen peroxide. The high temperature of combustion can encourage the formation of toxic oxides of nitrogen from the air used to burn the fuel:

$$N_2(g) + O_2(g) \rightarrow 2NO(g)$$

followed, in the atmosphere by:

$$2NO(g) + O_2(g) \rightarrow 2NO_2(g)$$

which contributes to the formation of acid rain, both in its own right, as nitric acid, and by oxidising the pollutant sulphur dioxide (from other sources) to sulphuric acid.

Questions

1   (a) Write the structural formulae for the following:

(i) 2-methylpropane;   (ii) 2,2-dimethylpropane;

(iii) 2,2,3-trimethylbutane.

(b) Give the systematic names for the following:

(i) $CH_3CH_2CH_2CH_3$;   (ii) $C(CH_3)_4$;   (iii) $CH_3CH_2CH(CH_2CH_3)_2$.

2   (a) Write down the structural formulae for all the isomers of C_6H_{14}.

(b) Write the systematic name for each isomer.

(c) Arrange the isomers as far as you can in order of increasing boiling points.

3   (This hard question requires some mathematical skill.) The enthalpy changes of combustion, ΔH_c, for methane, ethane and propane are –890, –1560 and –2220 kJ mol^{-1}. Derive an approximate expression for the enthalpy change of combustion of the hydrocarbon C_nH_{2n+2}. Use your expression to evaluate the enthalpy change of combustion of octane, C_8H_{18}. The listed value is –5510 kJ mol^{-1}. Comment on your answer.

4   The enthalpy change of combustion of propane is –2220 kJ mol^{-1}. The density of liquid propane is 0.58 g cm^{-1}. Calculate the calorific value of propane in kJ g^{-1} and in kJ cm^{-3}.

5   Give the equation, the conditions used and the names of the products when:

(a) but-2-ene reacts with hydrogen bromide;

(b) propene reacts with bromine.

Would you expect a species which attacked alkenes to be an electrophile or a nucleophile? Give a reason for your choice.

ALKANES AND ALKENES

6 (a) Write down the structural formulae of all isomers of but-2-ene and name them.

(b) Explain briefly the reason for the existence of these isomers.

(c) How would you show that each of these isomers was unsaturated?

7 Use the data from Table 3.2 on page 30 to plot graphs of

(a) boiling point against relative molecular mass

(b) melting point against relative molecular mass.

Comment on the graphs obtained in each case.

8 (a) Explain briefly why alkene molecules react with more substances than alkanes.

(b) From your knowledge of poly(ethene), deduce a possible structure for

(i) poly(propene),

(ii) poly(chloroethene), given that the formula of chloroethene is $CH_2=CHCl$.

Haloalkanes

Haloalkanes

Compounds formed when a member of the halogen group is substituted into an alkane are called haloalkanes. Haloalkanes are also referred to as halogenoalkanes.

General formula

$$C_nH_{2n+1}X$$

X is usually chlorine, bromine or iodine.

Members and nomenclature

The names of haloalkanes are derived from the name of the parent alkane. The name of the halogen present comes first, followed by the name of the alkane chain into which the halogen has been substituted. A number preceding the halogen indicates its position on the carbon chain. Some examples are:

CH_3Cl	chloromethane
CH_3CH_2Br	bromoethane
$CH_3CH_2CH_2Br$	1-bromopropane
$CH_3CHBrCH_3$	2-bromopropane

Functional groups

When atoms or groups of atoms other than carbon or hydrogen are present in an organic molecule, they are generally much more reactive than the hydrocarbon chain, which can only react in the ways already described for alkanes. Such groups are known as **functional groups** since it is these groups which determine the reactions of the molecule. The functional group in haloalkanes is the halogen atom. The carbon–halogen bond is usually easier to break than a carbon–hydrogen bond in the hydrocarbon chain. As a result, all the reactions of haloalkanes are reactions of the halogen atom, as will be seen later.

> **DEFINITION**
>
> A functional group is an atom (e.g. Br), a group of atoms (e.g. –OH) or a structural feature (e.g. C=C) in a molecule which confers characteristic chemical properties not shown by an alkane.

Structural isomerism

Structural isomerism can occur in the usual ways by:
- moving the halogen atom to different positions on the carbon chain as in 1-bromopropane and 2-bromopropane above
- branching of the carbon chain in larger molecules, for example, 2-bromobutane $CH_3CH_2CHBrCH_3$ and 2-bromo-2-methylpropane $(CH_3)_2CBrCH_3$, which are both isomers of C_4H_9Br.

Types of haloalkane

The hydrocarbon chain to which a functional group is attached can exist in one of three possible forms. As a result, there are three different types of haloalkane; in fact, there are three different types of compound in any homologous series containing monovalent functional groups. These types are **primary**, **secondary** and **tertiary**.

HALOALKANES

QUESTION

Which of these cannot exist as a tertiary halide?
C_2H_5I, C_3H_7Br, C_4H_9F, $C_5H_{11}Cl$?

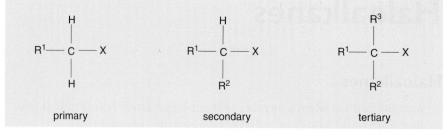

primary   secondary   tertiary

R^1, R^2 and R^3 are alkyl groups which may be the same or different but must contain at least one carbon atom.

QUESTION

Write and name the primary, secondary and tertiary forms of C_4H_9I.

Primary, secondary and tertiary are often abbreviated to 1°-, 2°- and 3°-, or in older books to p-, s- and t-.

A primary compound must have no fewer than two H atoms directly attached to the carbon atom that carries the functional group. They are therefore characterised by having the $-CH_2X$ grouping. Secondary compounds have only one H atom attached to the functional group carbon atom and are characterised by having a $-CHX$ grouping. Tertiary compounds have no H atoms on the functional group carbon atom.

The same types of reaction are open to all three types of haloalkane but there is a difference in the rate at which they react, as you will see in the next section.

In other homologous series, notably alcohols, the type of compound can be of greater significance and can influence the type of reaction, as well as the rate.

Rates of reaction of haloalkanes

Although all haloalkanes react with the same reagents and undergo the same types of reaction, the rate of reaction varies with two factors.

- **The nature of the halogen**. Iodides react more quickly than bromides, which in turn react more quickly than chlorides. This is because the iodine atom is bigger than bromine and so the carbon–iodine bond is longer than the carbon–bromine bond. Hence C–I has a lower bond energy than C–Br and is more easily broken. The similar, but opposite argument applies to the smaller chlorine atom in the shorter carbon–chlorine bond. The relevant values are shown in Table 4.1.
- **The type of haloalkane**. Tertiary compounds react more quickly than secondary, which in turn react more quickly than primary.

QUESTION

E (C–F) = 450 kJ mol^{-1}. How would you expect the activity of fluoroalkanes to compare with other haloalkanes?

Table 4.1 *Relationship between bond length and bond energy*

Bond	Bond length/nm	Bond energy/kJ mol^{-1}
C–Cl	0.177	338
C–Br	0.193	276
C–I	0.214	238

A tertiary iodide would be expected to react very much more quickly than a tertiary bromide and more quickly than a secondary iodide. Tertiary iodides would be the quickest of the common haloalkanes to react and primary chlorides the slowest. A practical method of demonstrating this will be outlined on p.51.

Reactions of haloalkanes

Substitution and elimination

Substitution reactions are those in which an atom or group is replaced by a different one

$$-C-X \rightarrow -C-Y$$

Elimination reactions are those in which atoms are lost from a molecule to form (usually) an unsaturated molecule and a simple molecule, often water or a hydrogen halide

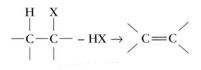

Because substitution occurs on the carbon atom, and because halogens are more electronegative than carbon, the carbon–halogen bond is always polarised as shown in Figure 4.1. The arrow on the bond suggests the displacement of the bonding electron pair.

$$\overset{\delta^+}{\underset{/}{\overset{\backslash}{C}}} \longrightarrow \overset{\delta^-}{X}$$

Fig. 4.1 Permanent polarisation of a bond.

Here substitution is brought about by nucleophiles like :OH⁻ which bring their own electron pair to form a new bond. The approach of these electrons to the carbon atom pushes the bonding C:X pair towards X; the ease with which this takes place depends on the polarisability of the C:X bond. This turns out to be more important than the permanent polarisation of the bond in controlling the rate of the reaction. Thus the longer, more diffuse C:I bond is easily polarised and iodoalkanes react rapidly, whereas the short, dense C:F bond is virtually inert.

In general, both substitution and elimination will occur, as well as some unexpected substitutions, but we can choose conditions which encourage one or the other.

Reaction with sodium or potassium hydroxides

Aqueous solutions *favour* substitution whereas solutions in ethanol *favour* elimination. Heat is required in both cases.

$$\text{NaOH(aq)}$$
$$CH_3CHBrCH_3 + OH^- \quad \rightarrow \quad CH_3CHOHCH_3 + Br^-$$

$$\text{KOH(ethanol)}$$
$$CH_3CHBrCH_3 + OH^- \quad \rightarrow \quad CH_3CH{=}CH_2 + Br^-$$

Do not imagine that in either case the reaction you are trying to favour will necessarily predominate. For some haloalkanes substitution is very difficult and for others elimination is hard to achieve.

HALOALKANES

A reaction in which a halogen is displaced by a hydroxyl group is often described as a **hydrolysis**.

In some cases it is possible to form two different alkenes in the same reaction. Consider, for example, the reaction of 2-bromobutane with 'ethanolic' potassium hydroxide. The Br atom must of course be removed from carbon atom number two, but there are two alternatives for the H atom. If it is removed from carbon atom number one, the product will be but-1-ene. If on the other hand, it is removed from carbon atom number three, the product is but-2-ene.

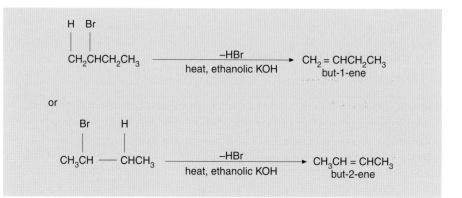

QUESTION

How would you expect ethylamine to react with iodomethane?

Reaction with ammonia

:OH^- has lone pairs of electrons on the O-atom which it can share to make covalent bonds and it is thus a nucleophile. Ammonia, :NH_3 also has an unshared pair and is similarly a nucleophile. It is not so strong a base as the hydroxide ion and is limited in its reactions to substitution (not elimination).

$$H_3N: + RI \rightarrow RNH_2 + HI$$
$$\text{primary amine}$$

Since ammonia is a base, for every molecule which reacts in this way one may be prevented from reacting by the acid produced

$$H_3N: + HI \rightarrow NH_4^+ \ I^-$$

The ammonium ion is no longer a nucleophile since the N-atom now shares all its valence (outer shell) electrons.

Thus a minimum of two moles of ammonia to every mole of haloalkane must be used.

$$2H_3N + RI \rightarrow RNH_2 + NH_4I$$

There is a second reason why an even larger excess of ammonia must be used. The amine product also has an unshared pair of electrons in the outer shell of N and, as soon as it is formed, it will compete with ammonia for the remaining haloalkane.

$$R\ddot{N}H_2 + RI \rightarrow R_2\ddot{N}H + HI$$
$$\text{secondary amine}$$

To minimise this side reaction a huge excess of (cheap) ammonia is used. The ammonia should be dissolved in an alcohol not water, or the equilibrium

$$NH_3 + H_2O \rightleftharpoons NH_4^+ + OH^-$$

will provide yet another competing nucleophile.

Because of the volatility of ammonia, coupled with the need to heat the reaction mixture (to give more energy to the molecular collisions) such reactions are usually carried out in a closed vessel under pressure.

Reaction with potassium cyanide

Hot ethanolic potassium cyanide reacts by substitution to give cyanides or nitriles.

The potassium cyanide is a source of cyanide ions, CN^-, and it is these which attack the haloalkane. Hence for iodoethane the reaction is:

$$CH_3CH_2I + CN^- \rightarrow CH_3CH_2CN + I^-$$
$$\text{propanenitrile}$$

It is beyond the scope of this unit to discuss the names or properties of nitriles but it is worthy of note that the nitrile produced contains one more carbon atom than the original haloalkane. This can be very useful in organic syntheses and will be dealt with more fully in Unit 4. The mechanism for these reactions will be given later.

Practical note. Many reactions use a process known as 'boiling under reflux'. This is a technique used frequently in organic chemistry where the reactions are often slow and the reagents volatile. The technique consists of using a condenser mounted vertically on top of the reaction vessel so that any vapours escaping during the heating process will condense to a liquid and run back into the flask. As a result the heating process can be carried out for a longer period.

Practical tests for halogen atoms

The halogen atoms in haloalkanes are covalently bonded to a carbon atom. In order to detect their presence, they are converted to the corresponding halide ion, which is then detected in the normal way, e.g. with silver nitrate solution.

This test can also be done by heating the haloalkane, *dissolved in ethanol*, with ethanolic silver nitrate solution. The same precipitates will occur. This method can also be used to test the reactivity of various haloalkanes by placing them in test tubes contained in a beaker of water maintained at constant temperature. The time taken for the precipitates to appear when silver nitrate solution is added indicates the rate of reaction. An alternative procedure is to heat a little of the haloalkane (RX) with aqueous sodium hydroxide for a few minutes to hydrolyse it:

$$RX + OH^- \rightarrow ROH + X^-$$

(At the end of this time the oily drops usually disappear if no alcohol was added to make the halide miscible at the start.) The solution is then acidified with nitric acid (to ensure that Ag_2O or $AgOH$ is not formed as a brown precipitate) and the halide ion is tested for with silver nitrate solution in the usual way.

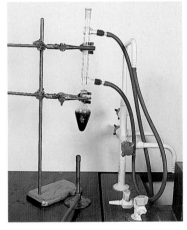

Fig. 4.2 Apparatus for heating under reflux.

QUESTION

Why is boiling under reflux NOT suitable for reactions of haloalkanes with ammonia?

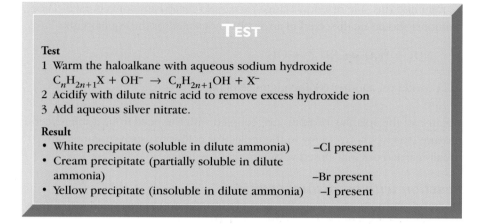

Test

1 Warm the haloalkane with aqueous sodium hydroxide

$$C_nH_{2n+1}X + OH^- \rightarrow C_nH_{2n+1}OH + X^-$$

2 Acidify with dilute nitric acid to remove excess hydroxide ion

3 Add aqueous silver nitrate.

Result

- White precipitate (soluble in dilute ammonia) —Cl present
- Cream precipitate (partially soluble in dilute ammonia) —Br present
- Yellow precipitate (insoluble in dilute ammonia) —I present

Haloalkanes and the environment

Many compounds with C–Cl bonds have been used as herbicides, e.g. the selective weedkillers 2,4-D and 2,4,5-T which farmers apply to grass crops and which make the countryside smell like TCP, a related product.

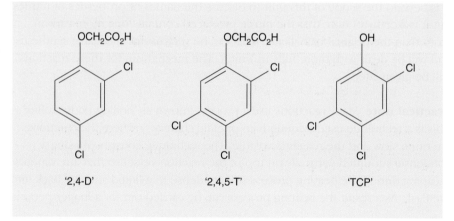

Fig. 4.3 Herbicides that resemble TCP.

Insecticides used include dieldrin, aldrin, DDT and lindane. The C–Cl bond is sufficiently inert to give these compounds a reasonably long life in all weather conditions but unfortunately their survival in the environment causes undesirable effects. Many, like DDT, are toxic to birds; insect-eating birds and those which prey upon them accumulate the toxic compound (or a breakdown product) in their body-fat with fatal results.

Questions

1   (a) Write down the structures of all isomers of C_4H_9Cl.

    (b) Assign a name to each isomer.

    (c) Classify each isomer as primary, secondary or tertiary.

2   Write equations and give the conditions for the reaction of an ethanolic solution of potassium hydroxide with:

    (a) 2-chloropropane;

    (b) 2-chlorobutane.

    Explain any differences in these reactions.

3   Write equations for the reactions of aqueous sodium hydroxide with:

    (a) 1-chloropropane;

    (b) 2-chloro-2-methylbutane.

4   A haloalkane is known to contain four carbon atoms. Describe simple experiments you would perform in order to show that:

    (a) the halogen present was bromine;

    (b) the molecule was saturated.

5   3-Bromo-3-methylhexane rapidly undergoes elimination when boiled with ethanolic potassium hydroxide. Write the structures of the FIVE alkenes which might be formed, and name them.

Functional groups containing oxygen

Alcohols

General formula

$$C_nH_{2n+1}OH$$

This could also be written as $C_nH_{2n+2}O$ but the formula above is more useful since it highlights the fact that one of the H atoms is different from the others in that it is bonded to an oxygen atom rather than a carbon atom. The functional group is clearly shown.

Formulae and nomenclature

The names of alcohols are simply based on the generic name of **alkanol**, as shown in Table 5.1. Isomerism occurs in the usual ways and a number is inserted before the **-ol** to indicate the position of the –OH group on the carbon chain when this is necessary.

Table 5.1 *The names of some alcohols*

Formula	Name
CH_3OH	methanol
CH_3CH_2OH	ethanol
$CH_3CH_2CH_2OH$	propan-1-ol
$CH_3CHOHCH_3$	propan-2-ol

The functional group and its test

The functional group is the –OH group, the presence of which can be shown by the test shown in the Test box.

> ### TEST
>
> To a sample of the alcohol in a clean, *dry* test tube, carefully add some solid phosphorus pentachloride (PCl_5). The result is evolution of 'steamy' acidic fumes of HCl.

The organic product of the reaction is always a haloalkane. The equation for the reaction in general is:

$$ROH + PCl_5 \rightarrow RCl + POCl_3 + HCl$$

and for the specific example of ethanol:

$$CH_3CH_2OH + PCl_5 \rightarrow CH_3CH_2Cl + POCl_3 + HCl$$
$$\text{chloroethane}$$

It is important that the test tube is dry, since water contains an –OH group and thus produces fumes of HCl which would invalidate the test.
Note that PCl_5 gives HCl with any –OH compound, e.g. ethanoic acid, CH_3COOH. It is not a specific test for alcohols only.

Some important reactions of alcohols

Just as haloalkanes can be hydrolysed to give alcohols, the –OH group of alcohols can be replaced by a halogen. The test reaction above is an example of this.
An alternative method is the use of a hydrogen halide

$$ROH + HX \rightleftharpoons RX + H_2O$$

In practice this is limited to the bromides – and then the hydrogen bromide is made *in situ*. The conditions are fairly drastic for hydrochloric acid which does not work well, and although hydriodic acid reacts well this is not a laboratory reagent since it is very rapidly oxidised in air. The hydrogen bromide is best made *in situ* by the action of concentrated sulphuric acid on potassium bromide:

$$KBr + H_2SO_4 \rightarrow KHSO_4 + HBr$$

Iodides are usually made by a variant of the phosphorus pentachloride method in which moist red phosphorus and iodine are heated with the alcohol.

$$2P + 3I_2 \rightarrow 2PI_3$$

$$PI_3 + 3ROH \rightarrow H_3PO_3 + 3RI$$

DEFINITION

in situ means in the place where the action is occurring.

Types of alcohol

As with the haloalkanes, there are three types:

- primary   containing CH_2OH
- secondary   containing $CHOH$
- tertiary   containing COH

Unlike the haloalkanes, the compounds can, in certain cases, behave differently towards a particular reagent (this is discussed later). All, however, give the positive test for the –OH group.

Dehydration

Most alcohols can be made to eliminate a molecule of water from their structures to give alkenes. This reaction is commonly referred to as **dehydration**. The usual methods are (i) heating the alcohol with a large excess of concentrated sulphuric acid, or (ii) passing the hot alcohol vapour over a heated catalyst such as aluminium oxide (vapour phase dehydration). Air must, of course, be excluded in the latter method.

The only requirement of the alcohol is that it should have a hydrogen atom on an α-carbon atom, i.e. the carbon atom next to that which carries the –OH group

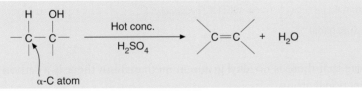

When there is more than one adjacent CH group then isomeric alkenes may arise:

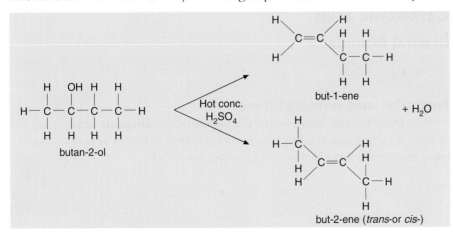

Aldehydes and ketones

These two homologous series are usually considered together since they have many reactions in common. They are only important, at this stage, as the oxidation products of alcohols.

General formulae

$C_nH_{2n}O$

Both groups have the same general formula but a more useful way of writing these shows the functional groups:

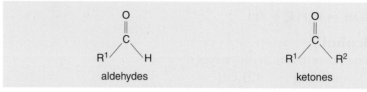

R[1] and R[2] are alkyl groups which may be the same or different. In the case of aldehydes, R[1] can also be a hydrogen atom but in the case of ketones, both groups must contain at least one carbon atom. Hence the simplest ketone contains three carbon atoms.

Formulae and nomenclature

The formulae and nomenclature follow from the generic names **alkanal** and **alkanone**, as shown in Table 5.2.

Table 5.2 *Names of aldehydes and ketones*

Aldehydes	
Formula	Name
HCHO	methanal
CH_3CHO	ethanal
CH_3CH_2CHO	propanal
$CH_3CH_2CH_2CHO$	butanal
$(CH_3)_2CHCHO$	2-methylpropanal

Ketones	
Formula	Name
CH_3COCH_3	propanone
$CH_3CH_2COCH_3$	butanone
$CH_3CH_2CH_2COCH_3$	pentan-2-one
$CH_3CH_2COCH_2CH_3$	pentan-3-one

Note that there is no alkyl group in methanal but there is a carbon atom in the functional group.

Carboxylic acids

General formula

$C_nH_{2n+1}CO_2H$

Formulae and nomenclature

Formulae and nomenclature follow the generic name **alkanoic acid**, as shown in Table 5.3. Note that the first acid has a value of $n = 0$. This is possible since there is a carbon atom in the $-CO_2H$ group.

Table 5.3 *The names of some carboxylic acids*

Formula	Name
HCO_2H	methanoic acid
CH_3CO_2H	ethanoic acid
$CH_3CH_2CO_2H$	propanoic acid
$CH_3CH_2CH_2CO_2H$	butanoic acid

The relationship between alcohols, aldehydes, ketones and carboxylic acids

These homologous series are linked together by a series of redox processes, as shown below.

Primary alcohols can be converted into aldehydes and carboxylic acids.

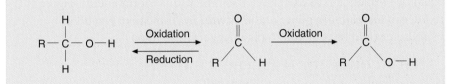

Secondary alcohols and ketones can be interconverted. Ketones resist further oxidation.

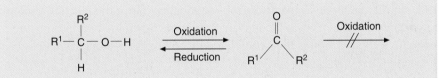

Tertiary alcohols resist oxidation.

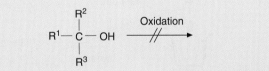

Knowledge of methods of reduction is not required for AS level.

Oxidation reactions

The oxidations shown above can be brought about by a number of oxidising agents. The one which is preferred is sodium or potassium dichromate, acidified with dilute sulphuric acid. This gives an easily observed colour change from orange to green on reduction from dichromate(VI) to chromium(III).

Differentiation between the different types of alcohol

These oxidation reactions can be used to distinguish between the different types of alcohol because the aldehydes or ketones produced can be recognised by the tests given earlier. Thus if a neutral liquid

(a) gives 'steamy' fumes with phosphorus pentachloride, it is probably an
    alcohol (or it may just be wet!)

INFORMATION

Primary alcohols, secondary alcohols and aldehydes are easily oxidised. Tertiary alcohols and ketones are resistant to oxidation.

QUESTION

Name and write the structures of all alcohols with molecular formula $C_4H_{10}O$. Predict the behaviour of each with potassium dichromate(VI).

(b) reduces potassium dichromate(VI), in dilute sulphuric acid it is probably primary (1°) or secondary (2°).

If the product of oxidation is an aldehyde or an acid, the alcohol is primary, and if the product is a ketone, the alcohol is secondary.

Preparation of an aldehyde

When oxidising a primary alcohol, it would be difficult to stop at the aldehyde stage were it not for the fact that the lower aldehydes are volatile and can escape before further oxidation occurs. A typical method for the preparation of ethanal would be slowly to add a mixture of ethanol and concentrated sulphuric acid from a separating funnel (which would replace the thermometer in Figure 5.1) to aqueous potassium dichromate(VI) which has been heated initially. The heat generated would maintain the temperature and drive out the aldehyde as it formed before further significant oxidation could occur. It would be very dangerous to mix the reactants in bulk at the start. The apparatus would be cleaned and reassembled, this time with a thermometer (Figure 5.1) in order to increase the purity of the ethanal by distillation. Fractional distillation would be required to obtain a good specimen.

$$CH_3CH_2OH \; + \; [O] \; \rightarrow \; CH_3CHO + H_2O$$
$$\text{ethanol} \qquad\qquad\qquad \text{ethanal}$$

A primary alcohol can be converted directly to a carboxylic acid in one step by heating under reflux with excess potassium dichromate(VI) in dilute sulphuric acid. As with aldehyde preparation, the potential vigour of the reaction usually requires the alcohol to be added in small amounts down the condenser (fitted for reflux); the reaction is then completed by boiling for a short while before distillation.

QUESTION

Why would you expect an aldehyde or a ketone to be more volatile than the corresponding (parent) alcohol?

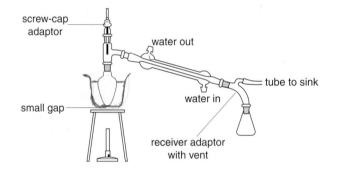

Fig. 5.1 *Distillation apparatus for the preparation of ethanal from ethanol.*

Synthetic pathways

Several functional groups have been studied to some extent and it is important to realise that the reactions of these functional groups are important for at least two reasons. Firstly, the reactions of the functional groups are always assumed to be the same whether they occur in simple molecules or in more complicated ones. They are also assumed to be the same when there are several functional groups within the same molecule. Secondly, in any reaction of a functional group, a product is formed. Hence the reaction provides a means of making or '**synthesising**' the product molecule. The reaction may not work very well in practice but it is at least a possible method of synthesis. The product molecule so formed will be capable of conversion into other molecules, and so on. Hence a series of reactions may be built up to convert one functional group into another. This is called a **synthetic route** or **pathway**.

A summary of synthetic pathways in this unit can be found in Figure 5.2.

The conversion of one organic molecule into another may be a simple one-step process or it may involve many steps.

For example, the conversion of ethanol into ethanal is a one-step process achieved by heating the ethanol with potassium dichromate(VI) and aqueous sulphuric acid:

$$CH_3CH_2OH + [O] \rightarrow CH_3CHO + H_2O$$

[O] is an old-fashioned way of indicating that (one atom of) oxygen has been added from an oxidising agent.

The conversion of 1-bromopropane to propanoic acid, however, involves two steps:

$$\begin{array}{ccc} & \text{step 1} & \text{step 2} \\ CH_3CH_2CH_2Br & \rightarrow & CH_3CH_2CH_2OH & \rightarrow & CH_3CH_2COOH \end{array}$$

Step 1 is achieved by boiling under reflux with aqueous sodium hydroxide. Step 2 is achieved by boiling under reflux with acidified potassium dichromate(VI) (see p. 58).

This diagram must be interpreted with a little care. Sometimes, on general diagrams of this type, it is not possible to carry out steps with compounds

QUESTION
Why must excess potassium dichromate be used?

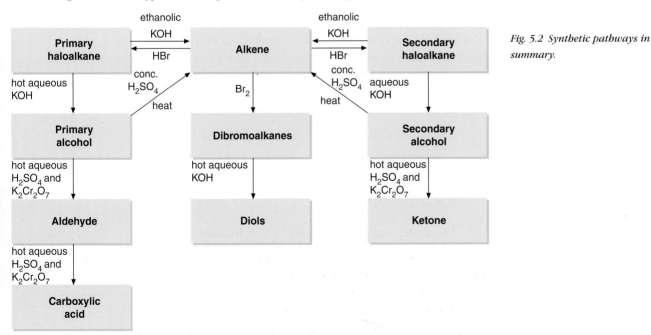

Fig. 5.2 Synthetic pathways in summary.

containing only one or two carbon atoms. Thus a primary haloalkane like iodomethane, CH_3I, cannot give an alkene on treatment with alkali since the simplest alkene has two carbon atoms per molecule. Similarly, only alkenes with three or more carbon atoms per molecule can give rise to secondary haloalkanes.

Calculating theoretical and percentage yields

Theoretical yield

Most organic preparations are designed to use a chosen material so that it is completely consumed. This may be because the material is difficult to remove unchanged from the product or (more often) because it is the most expensive material in the reaction mixture. When you make bromoethane from ethanol, potassium bromide, concentrated sulphuric acid (and a little water), all of the potassium bromide is used because the other compounds are in excess. In such a reaction the theoretical yield must be based on the amount of potassium bromide.

Example 1

Calculate the theoretical yield of bromoethane from 12 g of potassium bromide. Two methods are commonly used – they are arithmetically identical.

Method A: Write down the equation with the relevant masses in the equation as written (number of molecules in equation $\times$ molecular mass) underneath:

$$C_2H_5OH + KBr + H_2SO_4 \rightarrow C_2H_5Br + KHSO_4 + H_2O$$
$$119 \qquad\qquad \rightarrow \quad 109$$

In proportion: $\quad$ 12 g $\qquad \rightarrow \quad \dfrac{109}{119} \times 12\,g = 11\,g$

The answer is given to 2 significant figures because the mass of KBr was only given to 2 significant figures. Had the mass KBr been quoted as 12.0 g then it would have been correct to give the theoretical yield as 11.0 g.

Method B: Calculate the amount (in moles) of KBr, and hence the theoretical amount of bromoethane (same number of moles) and convert this to a mass:

$M_r(KBr) = 119\,g\,mol^{-1}$

Amount of KBr $= \dfrac{12\,g}{} = 0.101\,mol$

Amount of $C_2H_5Br = 0.101\,mol$

$M_r(C_2H_5Br) = 109\,g\,mol^{-1}$

Theoretical yield $= 109\,g\,mol^{-1} \times 0.101\,mol = 11\,g$

Percentage yield

This is used to measure the efficiency of a practical preparation:

$$\text{The percentage (\%) yield} = \frac{\text{Yield obtained} \times 100}{\text{Theoretical yield}}$$

Thus, in the above example, starting with 12 g of potassium bromide, if you obtained say 5.0 g of the product, bromoethane, then:

$$\text{The percentage (\%) yield} = \frac{5.0 \times 100}{11} = 45$$

More complicated examples

Suppose that you wanted to make 1-chlorobutane from butan-1-ol using phosphorus pentachloride. A handy textbook suggests that you use 10 g of butan-1-ol and 20 g of phosphorus pentachloride. In order to calculate the theoretical yield *you* have to find which of the two reactants to base it on:

$$C_4H_9OH + PCl_5 \rightarrow C_4H_9Cl + POCl_3 + HCl$$

$$M_r(C_4H_9OH) = 74 \text{ g mol}^{-1}$$

$$10 \text{ g } C_4H_9OH = 10 \text{ g}/74 \text{ g mol}^{-1} = 0.135 \text{ mol}$$

$$M_r(PCl_5) = 208.5 \text{ g mol}^{-1}$$

$$20 \text{ g } PCl_5 = 20 \text{ g}/208.5 \text{ g mol}^{-1} = 0.096 \text{ mol}$$

It can be seen that although the equation requires equimolar quantities of the reactants, the butanol is in excess. (This is not a prerequisite of the preparation – it is just what the 'handy textbook' might suggest.) The theoretical yield should thus be based on the 20 g of phosphorus pentachloride. The calculation is then as in the simple *Example 1 (Method B is perhaps easier as you have already worked out the amounts in moles)*.

Often organic reactants are in the liquid state and it is convenient to measure them out by volume. If the 'handy textbook' had suggested using 10 cm^3 of butan-l-ol, then you would have had to know and use the density of this liquid:

$$\text{Density of butan-l-ol} = 0.81 \text{ g cm}^{-3}$$

$$\text{Mass of butan-l-ol} = 10 \text{ cm}^3 \times 0.81 \text{ g cm}^{-3} = 8.1 \text{ g}$$

$$\text{Amount of butan-l-ol} = 8.1 \text{ g}/74 \text{ g mol}^{-1} = 0.109 \text{ mol}$$

The calculation of theoretical yield would still have to be based on the mass of phosphorus pentachloride.

Questions

1 For each of the homologous series alcohols, aldehydes, ketones and carboxylic acids:

(a) write the general formula for the series;

(b) give the formulae and the systematic names of the first three members of each series;

(c) give the functional group present in each series;

2 (a) Write down the structural formulae of all isomers of $C_4H_{10}O$.

(b) Name each isomer.

3 The following names were given by a student to five structures. They lead to the correct structure when drawn but they are not the correct names. Draw each structure in turn and then rename it correctly:
(i) 2-ethylpropane; (ii) hex-4-ene; (iii) 2-ethylpentan-5-ol;
(iv) 1-methylbutane; (v) 2,4-diethylpentane.

4 An alkene, A, on treatment with hydriodic acid gave a mixture of iodoalkanes, B and C, both containing 74.7% of iodine. B reacted with boiling aqueous potassium hydroxide to give two organic products; one, D, contained oxygen and one did not. C similarly reacted with boiling aqueous potassium hydroxide to give two organic products; one, E, contained oxygen and one did not. The oxygen-free product was the same in both cases. D and E both turned hot acidified potassium dichromate(VI) green; D gave a neutral oxidation product but that from E was acidic.

(a) Write down the structures of compounds A to E and name them.

(b) What was the other product formed at the same time as D and E?

(c) Which of B and C would you expect to be the major product? Give your reasons.

5 Predict the alkene products, if any, obtained by heating the following alcohols with concentrated sulphuric acid: (i) propan-l-ol; (ii) propan-2-ol; (iii) 2-methylpropan-l-ol; (iv) 2-methylpropan-2-ol; (v) methanol; (vi) butan-l-ol; (vii) 2,2,4,4-tetramethylpentan-3-ol.

Could any of the products exist as geometrical isomers?

6 An alcohol, $C_7H_{15}OH$, gives rise to three structurally isomeric alkenes on dehydration. Two of these alkenes exist as pairs of geometrical isomers. Name and draw the structure of the alcohol and draw the structures of the five alkenes, indicating the pairs of geometrical isomers.

7 4.6 g of ethanol on oxidation yielded 4.6 g of ethanoic acid. Calculate the percentage yield with respect to the ethanol. Express your answer to an appropriate number of significant figures.

8 20 g of l-iodobutane was heated with an aqueous solution of 10 g of potassium hydroxide. After distillation and purification, 5.0 g of butan-l-ol was obtained. Calculate the percentage yield of the alcohol and explain why this must be based on the initial quantity of the iodobutane and not on the potassium hydroxide.

Kinetics – how fast do reactions go?

Fig. 6.1 Erosion of a stone lion outside Leeds town hall. An example of a slow reaction, in which acid rain, formed by sulphur dioxide and nitrogen dioxide dissolving in rainwater, reacts with the calcium carbonate of which limestone is largely composed, converting it into more soluble calcium sulphate, which is then gradually washed away.

A few simple test tube reactions demonstrate that reactions proceed at very different rates. Some, like the reaction between hydrochloric acid and sodium hydroxide solutions, are extremely fast. Others, like the reaction between a lump of calcium carbonate and dilute hydrochloric acid, will bubble for some time before the production of carbon dioxide gas ceases. Many reactions between organic compounds are very slow indeed and require heating for a considerable time. You have already met examples of such reactions. It is the object of this chapter to consider some of the factors which affect the rate of a chemical reaction and to consider how useful a study of the kinetics of a reaction can be to our understanding of the way in which chemical reactions occur.

Factors which affect the rates of chemical reactions

Six factors commonly affect the rate of a chemical reaction:
- **concentration** of reactants in solution
- **pressure** of any gases present
- **surface area** of any solid reactants
- **temperature**
- **catalysts**
- **light** (occasionally)

An increase in pressure, temperature or surface area of a solid will lead to an increase in the rate of reaction. This is also true of the concentration of most reactants – though not all, as we shall see later.

Catalysts are substances which are capable of increasing the rate of a reaction without being chemically changed themselves. They will be dealt with later in this chapter.

Light only affects the rates of certain reactions, e.g. the reaction of chlorine with alkanes (see Chapter 3). The photographic industry largely depends on the effect of light on silver halides.

Theories of reaction rates

Theories about the speed of reactions must be able to explain how the factors mentioned above affect the rate of reaction in the way they do. There are two main theories of kinetics, the **collision theory** and the **transition state** theory.

The collision theory
The collision theory is based on the simple concept that before two particles can react they must collide. Only a small fraction of the total number of collisions, however, results in a reaction. There are two main reasons for this:
- the molecules must approach each other in the correct orientation (this is sometimes called the **steric factor**)
- the molecules must have a certain minimum amount of energy (often referred to as the **activation energy**).

Fig. 6.2 A much quicker reaction – exploding fireworks.

Direction of approach and orientation are both important steric effects. Two molecules travelling in much the same direction with high kinetic energy may not produce a high-energy collision; it is the energy of the collision which is important.

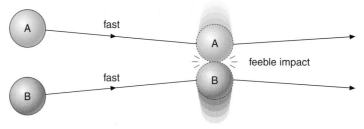

Fig. 6.3   Wrong direction.

However, if the molecules have energy below a certain minimum value, even a head-on collision may not be sufficiently violent to start the necessary bond-breaking.

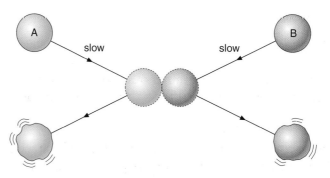

Fig. 6.4   Molecules have too little kinetic energy.

Suppose a large organic molecule with a small functional group is to be attacked by a reagent. If the bulk of the molecule protects the functional group from attack in a high-energy collision then a reaction may not occur.

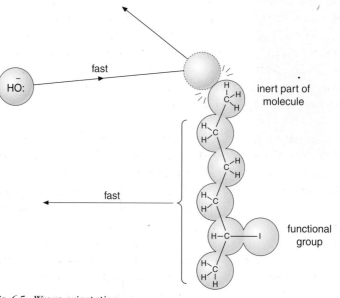

Fig. 6.5   Wrong orientation.

Only those molecules which collide in the right way with the right amount of energy will react.

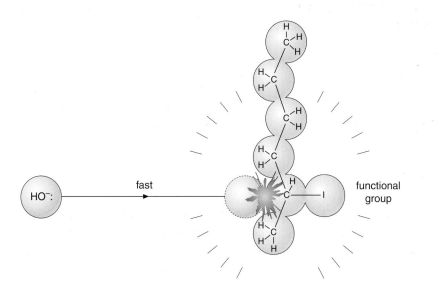

Fig. 6.6 *Effective collision.*

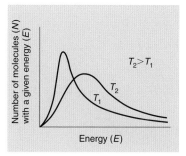

Fig. 6.7a *The Maxwell–Boltzmann distributions of molecular energy in a sample of gas at two different temperatures T_2 and T_1.*

Reaction rates can therefore be increased if collisions occur more frequently and/or the proportion of molecules having the required activation energy can be increased.

Thus if the concentration or pressure is increased then there are more particles in a given volume and they are bound to collide more frequently. Hence the increase in rate is explained simply by the increase in the number of collisions. Similarly, an increase in the surface area of a solid would lead to more collisions between the solid surface and the other reactant and hence the rate would increase.

In the case of an increase in temperature, the explanation is a little more complex since the kinetic energies of the molecules are increased along with the frequency of collisions. The former is by far the greater effect, however, and results in more molecules having energies greater than the minimum energy required. There are more molecules with sufficient energy to react *and* they collide more frequently. Hence the rate of reaction increases as there is an increase in the number of *effective* collisions per second.

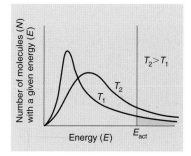

Fig. 6.7b *The population of molecules with $E \geqslant E_{act}$ at the lower temperatures T_1.*

The molecules in a gas (or a liquid) do not all have the same kinetic energies since they do not all have the same speeds. The way in which their energies are distributed is called the **Maxwell–Boltzmann distribution**. Figure 6.7 shows this distribution for a given sample of gas at two different temperatures T_1 and T_2, where $T_2 > T_1$. From this it can be seen that at any temperature, only a few molecules have very low or very high energies, most being around the most common value, represented by the peak value. As the temperature increases, the curve broadens out. The peak value decreases and moves towards a higher energy value. The area under the curve represents the total number of molecules in the sample and is therefore constant. The area under the curve beyond the activation energy E_{act} represents the number of molecules having energies greater than or equal to the activation energy. This increases as temperature increases and so therefore does the rate of reaction.

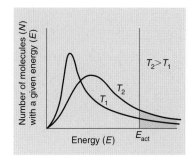

Fig. 6.7c *The population of molecules with $E \geqslant E_{act}$ at the higher temperature T_2.*

For the fast, enzyme-catalysed decomposition of hydrogen peroxide (H_2O_2) into water and oxygen at room temperature, the fraction of molecules with $E > E_{act}$ is about 10^{-4}. For the uncatalysed decomposition of the compound which is so slow that it needs to be heated in order to be obvious, the fraction of molecules with $E > E_{act}$ is a minute 10^{-14}. For most of the reactions you perform which have measurable speeds, the small fraction of viable molecules lies in this range. Therefore, for *practical* purposes, when you show this on a Maxwell–Boltzmann diagram, the E_{act} vertical line should be well to the right of the 'hump'.

The transition state theory

This theory considers the details of the actual collision between two molecules. As two molecules approach each other, repulsion between their electron clouds will push them apart again unless they have sufficient kinetic energy to overcome this repulsion. If they do get sufficiently close to each other, a rearrangement of electron clouds will occur so that some bonds are broken and new bonds form. While this is happening, a highly unstable species is formed for a very short period of time in which some bonds are partially broken and others partially formed. This unstable species is known as the **transition state** or **activated complex**. During this process, the kinetic energy of the collision is converted into potential energy, which can be shown on an enthalpy diagram usually, referred to as the **reaction profile** (Figures 6.8 and 6.9). The activated complex occurs at the peak of this profile and the energy gap between the reactants and this peak is known as the **activation energy E_{act}** for the reaction. Molecules must have this activation energy, or greater, when they collide if they are to react successfully, since this amount of energy must be absorbed even though energy is given out on going from transition state to products. Thus the activation energy can be considered as a barrier to reaction and the greater its value the slower the reaction will be. An example of a reaction involving a transition state is the substitution reaction of a primary haloalkane with aqueous alkali.

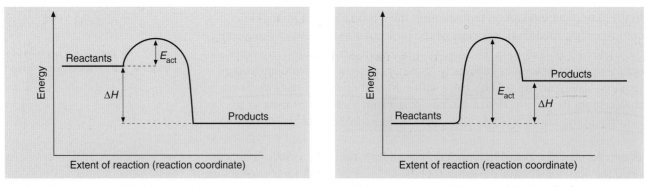

Fig. 6.8 *The reaction profile for an exothermic reaction.*   Fig. 6.9 *The reaction profile for an endothermic reaction.*

You can think of the hydroxide ion approaching the haloalkane, RCH_2X, using one of the unshared pairs of electrons on the O to displace the bonding pair between C and X:

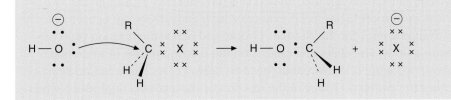

Fig. 6.10   Nucleophilic substitution.

The transition state will be the complex species when the O–C bond is half-formed and the C–X bond is half-broken (see Figure 6.11).

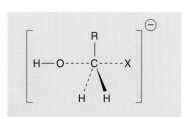

Fig. 6.11   A transition state.

This is an oversimplification because the ionic species in this reaction will all be stabilised by association with the solvent molecules.

Thermodynamic and kinetic stability

Reference was made in Chapter 1 to thermodynamic stability, which is loosely (see p.4) determined by the enthalpy levels of the reactants and products of a reaction. In a very exothermic reaction, the products are considered to be more stable than the reactants and the reactants are therefore classified as being thermodynamically less stable. We would normally expect that such a reaction would be spontaneous, i.e. likely to occur in the forward direction. This is in fact not always true, free energy change needs to be taken into account in determining whether or not a reaction is spontaneous (see Chapter 1). We shall, however, adopt the simplification that a negative ΔH is one of the principal driving forces of a reaction and so indicates the direction in which the reaction is likely to take place.

One reason why we do this is that we know how to calculate approximate enthalpy changes for the reaction of covalent substances from tabulated bond energies (see Chapter 3) and these are related to enthalpy changes, not to free energy.

However, such reactions frequently do not occur in practice and this is often due to the fact that the activation energy for the reaction is too high under the conditions being used. When reactants fail to undergo thermodynamically feasible reactions, with a high activation energy, they are said to be kinetically stable. An example of such a reaction is

$$C(s, graphite) + O_2(g) \rightarrow CO_2(g) \qquad \Delta H = -393 \, kJ \, mol^{-1}$$

Graphite in oxygen is *thermodynamically unstable* with respect to carbon dioxide and the reaction would be expected to be spontaneous from left to right. In practice, no reaction occurs at room temperature because the reactants are *kinetically stable* as the reaction has a high activation energy. At an elevated temperature the reaction proceeds as expected since the reactants now have sufficient energy to overcome the activation energy.

Catalysts

Catalysts are substances which increase the rate of a chemical reaction whilst remaining chemically unchanged themselves. They usually work by providing

more effective collisions

an alternative route for the reaction which has a lower activation energy than the normal route. Thus more molecules have enough energy to overcome the activation energy for the alternative route and hence the reaction proceeds more quickly. The effect of a catalyst on the reaction profile of a reaction is shown in Figure 6.12. Note that the introduction of a catalyst has no effect on either the enthalpy change of the reaction or the amounts of products when the reaction is completed.

> ### DEFINITION
>
> **Catalysts** are substances which increase the rate of a chemical reaction while remaining chemically unchanged themselves.

Frequently the catalyst may allow the reaction to proceed in two or more steps. For example, the reaction between peroxodisulphate ions and iodide ions:

$$S_2O_8^{2-}(aq) + 2I^-(aq) \rightarrow 2SO_4^{2-}(aq) + I_2(aq)$$

is catalysed by Fe^{2+} ions. The catalyst is believed to function by allowing two steps to occur thus:

$$S_2O_8^{2-}(aq) + 2Fe^{2+}(aq) \rightarrow 2SO_4^{2-}(aq) + 2Fe^{3+}(aq)$$

followed by:

$$2Fe^{3+}(aq) + 2I^-(aq) \rightarrow 2Fe^{2+}(aq) + I_2(aq)$$

Note that the catalyst (Fe^{2+} ions) is regenerated in the reaction and is not therefore used up. A possible reaction profile for this process is shown in Figure 6.13.

Fig. 6.12 The effect of a catalyst on the profile of a simple reaction.

Fig. 6.13 A possible reaction profile for a reaction catalysed by Fe^{2+} ions.

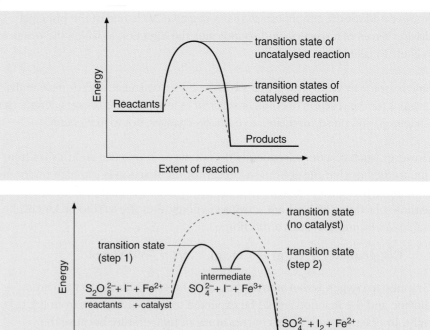

Most catalysed reactions, especially when catalysed by solids, have the form of two transition state 'humps' and an intermediate. They form such an intermediate when adsorbed on the surface of the catalyst.

Questions

1   Methane reacts with chlorine. Copper(II) oxide reacts with dilute sulphuric acid. Which of the following actions would increase the rate of reaction and why? (i) Raising the temperature; (ii) raising the pressure; (iii) shaking the container.

2   Explain the following two statements.

   (i)   the reaction of methane and oxygen has a large negative enthalpy change, yet it does not take place at 10 °C or 20 °C.

   (ii) If a tiny electric spark is produced in a mixture of methane and oxygen at 10 °C, the heat transferred is NOT sufficient to raise the temperature of the mixture by even 1 °C yet the reaction occurs with explosive violence.

3   Suggest why the neutralisation of hydrochloric acid by sodium hydroxide solution is virtually instantaneous.

7

Equilibria – how far do reactions go?

Reversible reactions

In studying chemical reactions so far, particularly in carrying out calculations, it has been assumed that the reactions go to completion. By this we mean that the reactants which are not in excess are totally used up and converted into products. For example, in calculating the mass of calcium oxide obtainable from heating 1.00 g of calcium carbonate at 900 °C, it must be assumed that the entire 1.00 g decomposes according to the equation:

$$CaCO_3(s) \rightarrow CaO(s) + CO_2(g)$$

and hence arrive at the answer of 0.56 g CaO.

If this reaction was performed in an open container this would in fact be the case. If, however, it was performed in a closed container, only part of the calcium carbonate would decompose, no matter how long the temperature was maintained at 900 °C. At this stage, the vessel would contain a certain amount of each of the three substances, which would not change provided that the temperature remained constant.

The reason for this is that the reaction is **reversible**, that is, the calcium oxide and the carbon dioxide also react together at this temperature to form calcium carbonate. Thus there are two reactions going on in opposite directions; obviously if the carbon dioxide were allowed to escape, the reverse reaction could not occur. It is standard procedure in chemistry to regard the substances on the left-hand side of an equation to be the reactants and those on the right to be the products. In this situation, however, all substances present are really both reactants and products. Nevertheless we shall continue to refer to the substances on the left-hand side of the equation as the reactants. The reaction from left to right will be referred to as the forward reaction and the one from right to left as the reverse reaction.

The nature of a chemical equilibrium

In any reversible reaction of this kind (assuming that we start with the substances on the left-hand side at a given temperature) the rate of the reaction from left to right will start at a certain level but will decrease as the concentrations of the reactants decrease. The reverse reaction will not start until some products have been formed, and even then the rate will be slow since their concentrations will be small. However, the rate of the reverse reaction will increase as the concentrations of the substances on the right-hand side of the equation increase (see Figure 7.1). Thus a point will be reached when the forward and the reverse reactions will be occurring at the same rate. At this point, the concentrations of the substances will remain constant but the reactions have not stopped. The reaction is said to have reached a state of **equilibrium** but it is a **dynamic equilibrium,** represented by the symbol $\rightleftharpoons$.

QUESTION

Predict the likely effect, on the forward and backward rates of a gaseous reaction in equilibrium, of adding some of the product.

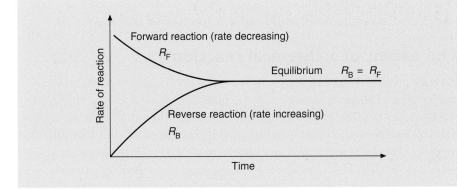

Fig. 7.1 Changes in the rates of forward and reverse reactions approaching equilibrium.

> ### DEFINITION
> A **dynamic equilibrium** is achieved when the forward reaction and the reverse reaction are occurring at equal rates and there are no concentration changes.

There are two types of equilibria: **homogeneous**, where all the components (reactants and products) are in the same phase, and **heterogeneous**, where the components are in different phases.

A phase, in many but not all systems, corresponds to a physical state. You should never assume that they are the same. It is a region in which there is no change in concentration of the substance(s) present and no boundary to cross. A pure liquid is one phase. A pure solution is also one phase; there will be at least two substances present (components) but no changes in concentration and, until you leave the solution, no boundary to cross. A saturated solution in a half-filled bottle in contact with air-and-vapour above it and undissolved crystals of solute at the bottom of the bottle would represent three phases. An emulsion, e.g. milk or 'emulsion paint' on the other hand, despite being all in the liquid state, is nevertheless, two phases because as you pass from the oily droplets to the aqueous solution dispersing them you must cross a boundary. A little water in a large evacuated container would also be two phases; although there is only one component (H_2O) you would pass a boundary as you moved from liquid (high concentration of H_2O) to vapour (at low concentration).

The thermal decomposition of calcium carbonate is an example of a heterogeneous system and should be written with the appropriate equilibrium sign:

$$CaCO_3(s) \rightleftharpoons CaO(s) + CO_2(g)$$

Some examples of homogeneous systems are:

$$CH_3CO_2H(l) + C_2H_5OH(l) \rightleftharpoons CH_3CO_2C_2H_5(l) + H_2O(l)$$

$$2SO_2(g) + O_2(g) \rightleftharpoons 2SO_3(g)$$

$$H_2(g) + I_2(g) \rightleftharpoons 2HI(g)$$

It should be noted that:
- a dynamic equilibrium will only be achieved in a closed system (Figure 7.2), that is, none of the components of the equilibrium can escape;

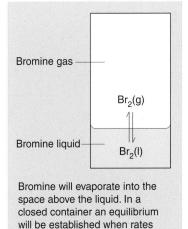

Bromine will evaporate into the space above the liquid. In a closed container an equilibrium will be established when rates of condensation and evaporation are equal.

Fig. 7.2 Dynamic equilibrium in a closed system.

• the same equilibrium will be achieved, provided the temperature is constant, no matter from which end it is approached (see Figure 7.1).

The extent of a chemical reaction

Having established the fact that many reactions do not go to completion but reach a state of dynamic equilibrium, the question arises as to how far the reaction goes before the equilibrium is established. The extent of a reaction when an equilibrium is established is called the **position of equilibrium**. This varies from one reaction to another and depends upon the temperature at which the reaction is performed. If a reaction uses more than 50% of the reactants before reaching equilibrium then the position of equilibrium is said to lie to the right. On the other hand, if less than 50% of reactants are consumed before the equilibrium is reached, the position of equilibrium is said to lie to the left. It must *not* be assumed that equilibrium is established when there is 50% of reactants and 50% of products.

Factors affecting the position of equilibrium

There are three factors which can be changed that may result in a change in the position of equilibrium. These are:

• concentration
• pressure
• temperature.

These three factors must be considered separately.

The effect of changes in concentration at constant temperature

Consider a system in **equilibrium** and represented by the equation:

$$n\text{A} + m\text{B} + x\text{C} \rightleftharpoons p\text{P} + y\text{Q}$$

The effect of adding more A, B, C, P or Q must be predicted with some care.

If you increase the concentration of A then some of the extra A will react with B and C to give more P and Q until equilibrium is restored. We say that the position of equilibrium moves to the right. In general, if a reaction is in equilibrium and you increase the concentration of a reactant then the position of equilibrium moves to the right; if you increase the concentration of a product then it moves to the left.

I have been extremely careful not to write 'if you add more A' because the argument only applies safely to concentration, not amount. Most of the time you can get away with saying 'if you add more...' but look at this possibility:

$$\text{A} + \text{B} \rightleftharpoons \text{C} + \text{D}$$

If you add more A and, as a result, increase the total volume, e.g. suppose A, B, C and D were gases at atmospheric pressure, then you will certainly increase the concentration of A but will reduce the concentrations of all the other species. It is possible (for teachers and examiners trying to prove a point) to show that if you add a little nitrogen to the equilibrium reaction for the formation of ammonia at constant pressure:

$$N_2(g) + 3H_2(g) \rightleftharpoons 2NH_3(g)$$

the consequent reduction in the concentration of hydrogen is inadequate to cope with the increased concentration of nitrogen. The result is that if, as suggested, you add a small amount of nitrogen, the reaction produces more nitrogen from the ammonia. Fortunately you have to look hard and far to find such examples.

The effect of increasing the overall pressure

Changing pressure only affects gaseous equilibria and, only then, if there is a change in the total number of molecules. If you increase the pressure on an equilibrium mixture then the position of equilibrium moves towards the side with the smaller number of molecules.

Increasing the pressure would move the positions of equilibrium of all the reactions below to the right (RHS)

$$2SO_2(g) + O_2(g) \rightleftharpoons 2SO_3(g); \quad \text{3 molecules} \rightleftharpoons \text{2 molecules}$$

$$N_2(g) + 3H_2(g) \rightleftharpoons 2NH_3(g); \quad \text{4 molecules} \rightleftharpoons \text{2 molecules}$$

$$2NO_2(g) \rightleftharpoons N_2O_4(g); \quad \text{2 molecules} \rightleftharpoons \text{1 molecule}$$

Even though they are gaseous, pressure would have no effect on the equilibria

$$H_2(g) + I_2(g) \rightleftharpoons 2HI(g)$$

$$H_2O(g) + CO(g) \rightleftharpoons H_2(g) + CO_2(g)$$

The effect of temperature change

Ignore the effect of temperature on the pressure of a gas for the purposes of this discussion; temperature and pressure effects must be clearly distinguished.

The positions of equilibrium of all reactions which involve an enthalpy change are affected by changing the temperature. If the temperature is raised (by supplying heat) the position of equilibrium moves in the endothermic direction (i.e. it absorbs some of the heat you supplied). Similarly, lowering the temperature moves the position of equilibrium in the exothermic direction.

The reaction below is important for the manufacture of hydrogen. The position of equilibrium is unaffected by pressure changes. It is endothermic and thus the formation of hydrogen is favoured by high temperatures.

$$H_2O(g) + CO(g) \rightleftharpoons H_2(g) + CO_2(g); \quad \Delta H = +41 \text{ kJ mol}^{-1}$$

These ideas were first unified in the nineteenth century. The way in which the position of equilibrium changes, if at all, was deduced by **Le Chatelier**.

> ## QUESTION
>
> Predict the effect of increasing the pressure on the position of equilibrium of the reactions:
> $$NO(g) + NO_2(g) \rightleftharpoons N_2O_3(g)$$
> $$Br_2(g) + Cl_2(g) \rightleftharpoons 2BrCl(g)$$

The rate of attainment of equilibrium

From the discussion of kinetics in Chapter 6 it will be apparent that the factors that affect the position of equilibrium will also affect the rate at which the equilibrium is reached. Thus an increase in concentration, pressure (if gases are involved) and temperature will all increase the rate at which the equilibrium is established, as well as having an effect on the position of equilibrium. The explanations for this, based on the collision theory, have already been given (see Chapter 6).

A catalyst will increase the rate at which an equilibrium is established but will have no effect on the position of the equilibrium. This is because in providing a new route for the reaction, lower activation energy routes are provided for both the forward and the reverse reactions. Hence the rates of both forward and back reactions increase, but the position of equilibrium remains unaltered. Reaction profiles for catalysed and uncatalysed reactions are shown in Figure 7.3.

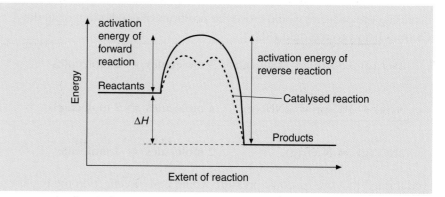

Fig. 7.3 Energy profile for a reversible reaction which is exothermic in the forward direction, showing the effect of a catalyst.

Questions

1. State whether the equilibrium position moves to the right, moves to the left or stays the same for each of the following systems, when the total pressure is increased. State your reasoning in each case.

   (a) $N_2(g) + 3H_2(g) \rightleftharpoons 2NH_3(g)$

   (b) $H_2(g) + I_2(g) \rightleftharpoons 2HI(g)$

   (c) $2SO_2(g) + O_2(g) \rightleftharpoons 2SO_3(g)$

   (d) $N_2O_4(g) \rightleftharpoons 2NO_2(g)$

2. For the equilibrium $2NO(g) + O_2(g) \rightleftharpoons 2NO_2(g)$, the enthalpy change for the forward reaction is negative. State and explain the effect on the equilibrium position of each of the following.

   (a) An increase in temperature at constant pressure.

   (b) A decrease in the concentration of oxygen.

   (c) The addition of a platinum catalyst.

   (d) An increase in the concentration of NO.

3 X is a product of a gaseous reaction which results in an equilibrium mixture being formed:

$$Reactants \rightleftharpoons X$$

The percentage of X in the equilibrium mixture at various temperatures and pressures is shown in the following table.

	1 Atm	100 Atm	200 Atm
550 °C	0.77	6.70	11.9
650 °C	0.032	3.02	5.71
750 °C	0.016	1.54	2.99
850 °C	0.009	0.87	1.68

Use this data to deduce, giving your reasoning in each case,

(a) whether the production of X is exothermic or endothermic. *exothermic*

(b) whether the production of X involves an increase or decrease in the number of moles of gas present. *decrease*

(c) the best conditions to obtain the greatest yield of X. *low temperature, high pressure*

4 For the equilibrium:

$$H_2(g) + I_2(g) \rightleftharpoons 2HI(g); \quad \Delta H = +52 \text{ kJ mol}^{-1}$$

what would be the effect, if any, on this equilibrium of:

(a) increasing the total pressure at constant temperature; ✗

(b) increasing the temperature at constant pressure? ⟶

Give your reasoning in each case.

5 Predict the effect of: (i) increasing the temperature, (ii) increasing the pressure, on the position of equilibrium of each of the following reactions.

(a) $N_2(g) + O_2(g) \rightleftharpoons 2NO(g)$ $\Delta H = +180 \text{ kJ mol}^{-1}$ ⟶ ✗

(b) $H^+(aq) + OH^-(aq) \rightleftharpoons H_2O(l)$ $\Delta H = -57 \text{ kJ mol}^{-1}$ ⟵ ✗

(c) $3O_2(g) \rightleftharpoons 2O_3(g)$ $\Delta H = +284 \text{ kJ mol}^{-1}$ ⟶ ⟶

6 Suggest why the following are true.

(a) Industrial catalysts (like platinum) are finely divided, and fine wire or mesh is preferred to powder. *increase the surface that can touch with the reactants*

(b) Although speed is important in industry, there is often quite a low upper limit to the temperature of industrial reactions. *give a better yield*

8

Industrial inorganic chemistry

Industrial applications

Many industrial processes involve an equilibrium at some point in the manufacturing process. One such is the **Haber–Bosch** process for the manufacture of ammonia.

The key step in this process is the direct synthesis of ammonia from nitrogen and hydrogen, in which an equilibrium is established.

$$N_2(g) + 3H_2(g) \rightleftharpoons 2NH_3(g) \qquad \Delta H = -92\,kJ\,mol^{-1}$$

Application of the principles of equilibrium

The formation of ammonia is exothermic and produces fewer molecules. Thus the position of equilibrium should be favoured by low temperatures and high pressures.

As can be seen from Table 8.1, we might predict that use of a pressure of $6.00 \times 10^4\,kPa$ and a temperature as low as 473 K would achieve almost complete conversion to ammonia.

Economic factors to be considered

There are, however, other factors which affect the economics of the process and which must be taken into account; these are:

- containers operating at high pressures are expensive to build and to operate
- a reduction in temperature slows down the rate at which the equilibrium is attained
- catalysts can be used in order to increase the rate of reaction
- catalysts are often susceptible to poisoning by impurities present, particularly sulphur compounds and carbon monoxide in the original feedstock
- catalysts usually last longer at lower temperatures
- unreacted gases can be recycled if product can be removed, thus allowing a continuous flow process.

In practice a compromise set of conditions is applied so that the requirement to push the equilibrium position to the right is balanced against the requirement to keep the rate of reaction at a reasonable level. The conditions used in many plants are a pressure of $2.00 \times 10^4\,kPa$ (200 atm) and a temperature of 673 K (400 °C) which keeps the rate of reaction at a reasonable level as well as giving a reasonably long life to the catalyst (about 5 years). An iron-based catalyst increases the rate of attainment of equilibrium and the incoming gases are purified before entering the catalyst chamber in order to avoid poisoning the catalyst.

Although a conversion of about 40% would be possible under these conditions (as shown in Table 8.1), the gases do not spend long enough in the catalyst chamber to reach equilibrium and a conversion of only about 15% is achieved. The gas mixture is cooled in order to liquefy the ammonia and remove it from

Table 8.1 *The percentage of nitrogen and hydrogen converted to ammonia for different conditions of temperature and pressure*

Pressure /10^2 kPa	Temperature/K			
	473	573	673	773
10	51	15	4	1
100	82	53	25	11
200	89	67	39	18
300	93	71	47	24
400	94	80	55	32
600	95	84	65	42

QUESTION

Ammonia boils at –33°C but the liquefaction plant uses cold water to remove the ammonia from the gas stream. Explain.

the remaining gases. The unreacted nitrogen and hydrogen are then passed through the catalyst beds again, thus maintaining a continuous circulation.

The raw materials for the Haber process

The hydrogen is obtained from natural gas, which contains methane, CH_4, or the naphtha fraction of petroleum, a hydrocarbon mixture by reaction with steam in the presence of a nickel catalyst:.

$$CH_4(g) + H_2O(g) \rightarrow CO(g) + 3H_2(g)$$

Carbon monoxide would poison the catalyst and must therefore be removed from the gases before passing over the catalyst. Air is injected into this mixture and, by a series of reactions, nitrogen and hydrogen are produced in a 1:3 ratio, as required by the process, and carbon monoxide, now oxidised to carbon dioxide, is removed from the system.

Although natural gas is readily available at the moment, it will not always be so. The alternative fossil fuel, coal, which was previously used (in the form of coke), could be used again, but this is also a finite resource. The equation for the generation of hydrogen would then be:

$$C(s) + H_2O(g) \rightarrow CO(g) + H_2(g)$$

The manufacture of nitric acid

One of the main uses of ammonia is in the manufacture of nitric acid. The main stage in this process is the oxidation of ammonia.

Despite its high hydrogen content, ammonia, unlike methane, CH_4, is not a flammable gas in air. It will burn in oxygen. The combustion of methane gives CO_2 and H_2O; both oxides have negative enthalpies of formation and are thermodynamically stable with respect to their elements. Ammonia, however, gives N_2 and H_2O; all the common oxides of nitrogen have positive enthalpies of formation and are less thermodynamically stable than the elements.

In the presence of a platinum–rhodium catalyst however, ammonia can be oxidised by air to give nitrogen monoxide:

$$4NH_3(g) + 5O_2(g) \rightarrow 4NO(g) + 6H_2O(g) \qquad \Delta H = -1636 \text{ kJ mol}^{-1}$$

QUESTION

Periodically, the gases in the Haber plant have to be 'purged' (cleared out) and fresh gas put in. What gas would build up in the plant?

The temperature of the catalyst must be carefully controlled. At too low a temperature the slower reaction may not go to completion; the presence of ammonia in the subsequent stages of nitric acid production would be disastrous. At too high a temperature some of the ammonia would be oxidised to nitrogen (and water) representing an uneconomic loss of ammonia. The temperature is adjusted to and maintained at about 900 °C by controlling the flow rate of the gases in this highly exothermic reaction.

The process is operated under increased pressure because this packs more reactants into the same capacity plant and increases the rate slightly by increasing the number of molecular collisions per second at the catalyst surface. An excess of air is used to ensure complete oxidation of ammonia.

Cold air is added to the mixture as it leaves the catalyst because the next stage is an exothermic equilibrium and is therefore favoured by lower temperatures:

$$2NO(g) + O_2(g) \rightleftharpoons 2NO_2(g) \qquad \Delta H = -94 \text{ kJ mol}^{-1}$$

Extensive cooling of the gases is necessary before this (exothermic) process can be brought to effective completion and before the resulting gases can be absorbed in cold water. The heat is used elsewhere in the plant, e.g. to preheat the ammonia and air before the catalyst stage.

Finally the nitrogen dioxide is absorbed in water in the presence of air to give nitric acid. This final reaction (sequence) is best summarised by the equation:

$$2H_2O(l) + 4NO_2(g) + O_2(g) \rightarrow 4HNO_3(aq)$$

Thus the actual conditions used, leading to about a 96% conversion, are:
- pressure: 4–10 atmospheres
- temperature: 975–1225 K
- catalyst: platinum containing 10% rhodium.

The manufacture of sulphuric acid by the Contact process

There are two essential steps in this process

$$2SO_2(g) + O_2(g) \rightleftharpoons 2SO_3(g); \Delta H = -200 \text{ kJ mol}^{-1}$$

and

$$H_2O(l) + SO_3(l) \rightarrow H_2SO_4(l); \Delta H = -130 \text{ kJ mol}^{-1}$$

(the enthalpy changes are stated for standard conditions when sulphur trioxide is liquid). The second reaction will be dealt with first before discussion of the equilibrium.

The most obvious characteristic of the second reaction is its extreme exothermicity. Reaction with water has therefore to be done under controlled conditions and is usually achieved by adding sulphur trioxide and the appropriate quantity of water, separately, to a large volume of concentrated sulphuric acid which acts as a diluent.

Application of our ideas about equilibrium to the first reaction suggests that the product would be favoured at equilibrium if high pressures were used (because three gaseous molecules form two gaseous molecules) and at low temperatures (because the reaction is exothermic).

Provided that the temperature is not too high, the equilibrium is a favourable one and little advantage is gained by using high pressures with all the attendant difficulties and cost. The use of an excess of (free) air at just sufficiently increased pressure to push the gases round the plant helps to ensure efficient oxidation.

Most industrial processes are exothermic and there is an inevitable competition between low rates with a favourable position of equilibrium at low temperatures, and rapid achievement of a less favourable position of equilibrium at high temperatures. The problem is solved in the usual way by using a catalyst at the lowest temperature that is consistent with a fast rate; in this case vanadium(V) oxide at about $440\,°C$ is used. The gases must be clean and dry to prevent catalyst poisoning and the catalyst must be finely divided on a support (powder would blow about) because the catalysis is heterogeneous and reaction takes place on the surface of the catalyst. Usually the rate of passage of the gas over the catalyst is such that the exothermic reaction maintains the temperature once the plant is operating. The vanadium(V) oxide is probably reduced by the sulphur dioxide and oxidised by the air alternately.

Modern plants pass the reacting gases through several catalyst beds, cooling the gases between each, thereby more nearly achieving ideal conditions. Thus, 98% conversion of the sulphur dioxide is possible. Present day environmental legislation does not allow the remaining SO_2 to be vented to the atmosphere. Various methods are used to avoid this. After removal of the SO_3 the small remnant of SO_2 and air may be preheated and then passed over a separate catalyst to convert perhaps 90% of what is left into SO_3 which can then be absorbed as before. Finally the partially deoxygenated air and trace of SO_2 may be 'scrubbed' with water before returning to the atmosphere.

The raw materials for the Contact process
The majority of sulphur dioxide is obtained by burning imported sulphur in air

$$S(l) + O_2(g) \rightarrow SO_2(g)$$

As heavy metal refining has decreased in the UK the roasting of (imported) sulphide ores has decreased in importance, e.g.:

$$2ZnS(s) + 3O_2(g) \rightarrow 2ZnO(s) + 2SO_2(g)$$

Ammonia, nitric acid and sulphuric acid in the fertiliser industry
Nitrogen is a key element in fertilisers and it is not surprising that ammonium nitrate (Nitram) is widely used in agriculture. It is produced by direct union of ammonia and (diluted) nitric acid

$$NH_3 + HNO_3 \rightarrow NH_4NO_3$$

A clean, water-soluble crystalline solid, its handling properties are only marred by its power as an oxidising agent. Mixed with organic material such as sawdust or diesel fuel, of which there may be plenty on a farm, it can be dangerously explosive.

Ammonium sulphate, which contains less nitrogen and is safer to store, is made from sulphuric acid but not normally directly with ammonia as it is the by-product of other reactions.

Sulphuric acid is also used to manufacture phosphate fertilisers. Natural calcium phosphate rock which, for the sake of discussion can be simplified to $Ca_3(PO_4)_2$, is too insoluble to give immediate benefit to crops. In one (older) process, stirring the material with the correct amount of moderately concentrated sulphuric acid gives the more readily soluble calcium dihydrogen phosphate

$$Ca_3(PO_4)_2(s) \; + \; 2H_2SO_4(aq) \; \rightarrow \; Ca(H_2PO_4)_2(s) \; + \; 2CaSO_4(s)$$

Most of the water in the mixture is taken up in the hydrated form of calcium sulphate which is left as part of the product since it does no harm. The mixture is sold as 'superphosphate'.

The extraction of aluminium

Uses of the metal

Aluminium is a metal of outstanding importance. Its low density makes it ideal for vehicle construction, although the mechanical weakness of the pure metal requires that it be alloyed to increase its strength; the coachwork or body of many vehicles, cars, buses and underground trains may be constructed from aluminium (alloy) strengthened by a chassis of steel. A better conductor of heat than iron, it vies with the cheaper but denser metal for car engine blocks. Its good electrical conductivity and low density allow it to compete successfully with the more expensive and better electrical conductor, copper, for the manufacture of electric cable; however, for overhead power lines it needs to be supported by a steel core to prevent it stretching and breaking.

Probably no use of aluminium depends on a single property. The well-known resistance of aluminium to corrosion makes it suitable for greenhouse frames, but our ability to extrude the metal in continuous bars of complicated cross-section (to increase the mechanical strength) is very important. The widespread use of aluminium for the complicated reflectors of modern car headlamps depends on both the high reflectivity of the polished metal and the ease with which sheets can be stamped into shape (malleability).

Other uses include the manufacture of kitchenware and food containers, wrapping foil and food containers for a diverse range of products, from beer to sardines, highly reflective coatings on glass in cameras and telescopes and as 'chaff' dropped from military aircraft to confuse radar systems.

Raw materials

Aluminium compounds are widely distributed in the Earth's crust. The bulk of the material is in the form of aluminosilicates (igneous rocks) and clays (the result of weathering of igneous rocks), which also contain a large proportion of combined silica. Unfortunately no-one has found an economic way of extracting the aluminium from these abundant sources, and the purer, white clays are already in great demand for ceramic and paper manufacture and are thus expensive. All the (primary) aluminium is extracted from the much less abundant hydrated aluminium oxides, $Al_2O_3.H_2O$ and $Al_2O_3.3H_2O$ or mixtures of these loosely described as 'bauxite'.

The high activity of aluminium or, more precisely, the high negative free energy of formation of its oxide, ΔG_f, make electrolysis of a melt the only feasible method of extraction.

Here the relative values of ΔH_f provide a satisfactory guide to stability of the oxides (and the consequent difficulty of regaining the metals), as seen by comparison with those for ΔG_f (see Chapter 1). The absolute values of these figures have to be treated with care. Both of these oxides have the form M_2O_3. Therefore the formula refers to the same amount (i.e. number of moles) of the metal or of oxygen. Direct comparison of data for say CaO and Al_2O_3 is not valid.

Table 8.2 *Free energy and enthalpy changes ($kJ\,mol^{-1}$) for the formation of the oxides of aluminium and iron*

	ΔG_f	ΔH_f
Al_2O_3	−1580	−1680
Fe_2O_3	−820	−740

Many metals are purified during the process of extraction and then, if necessary, further purified. Iron has many impurities removed in the slag of the blast furnace and the crude product may be cast directly or further purified during steel manufacture. Similarly, copper loses many of its impurities in a thermal reduction process but final purification is by electrolysis. The extraction of aluminium reverses this procedure; the raw material is purified before the extraction process. An outline knowledge of the purification of the ore is required for the Edexcel examination, but you are not required to know any equations for the purification; this is partly because the ones commonly used poorly represent the species which are formed.

DEFINITION

An amphoteric oxide is one which can react like a base in the presence of an acid and like an acid in the presence of a base. In both cases water and a salt are formed.

The crude oxide is crushed and heated under pressure with concentrated aqueous sodium hydroxide. The (amphoteric) aluminium oxide dissolves to give sodium aluminate solution:

$$2NaOH(aq) + Al_2O_3(s) \rightleftharpoons 2NaAlO_2(aq) + H_2O(l)$$

Some of the accompanying silica also dissolves because silica is acidic

$$2NaOH(aq) + SiO_2(s) \rightleftharpoons Na_2SiO_3(aq) + H_2O(l)$$

Iron(III) oxide does not react with the alkali because it is a basic oxide and it is left behind in a fine 'red oxide mud' which is used to manufacture protective paint for ironwork – you often see it on girders.

Freshly precipitated aluminium hydroxide rapidly changes into a very insoluble form (hydrated aluminium oxide). A little of this form is added to the cold, filtered aluminate solution to 'seed' it. The first reaction is then effectively reversed, and the precipitated hydrated aluminium oxide is filtered off, washed and roasted to give pure aluminium oxide used for the extraction of the metal. Some aluminium remains in the solution but this will be used again for the next extraction. The life of the alkali is limited, however, by the presence of increasing amounts of the silicate.

You should be aware that the transport and crushing of the bauxite, heating, and the use of large amounts of sodium hydroxide add considerably to the cost of the overall process during the conversion of the crude bauxite into anhydrous aluminium oxide.

Electrolytic extraction

Aluminium oxide is high melting (m. 2040°C). This renders it totally unsuitable as an electrolyte. Even if the electrical properties of the molten oxide were satisfactory, the problems of handling such an electrolyte would be formidable, for example, no steel components (m. ca. 1550°C) could be in contact with it.

The electrolyte is a bath of molten cryolite (Na_3AlF_6) containing about 10% calcium fluoride, into which up to 10% of aluminium oxide is dissolved. Cryolite is an uncommon mineral, and the bulk of it is manufactured. The mixture is both melted and maintained at its melting temperature, just below 1000°C, by the electric current of up to 100 000 amps. The cell operates at about 5 volts (DC), rising a little as each batch of aluminium oxide is exhausted. Only about 2 volts is required to decompose the oxide; the rest is required to overcome the electrical resistance. This results in the output of a huge amount of heat which corresponds to about 60% of the 500 kW or so supplied to each cell. The balance between heating and electrolysis is an important one and it is necessary to maintain the moveable anodes at a constant distance (about 5 cm) from the aluminium above the cathode. The need for so much heat, which not only keeps the electrolyte molten but also preheats each batch of new aluminium oxide, is the single most important factor in the cost of the process. The carbon electrodes must be very pure since the aluminium is rarely purified after production.

At the cathode, the cell lining being made of carbon blocks, aluminium is liberated (m. 660°C) and accumulates in a molten condition at the base of the cell.

$$Al^{3+} \text{ (solution)} + 3e^- \rightarrow Al(l)$$

When the aluminium oxide content falls to critical levels, the molten aluminim is rapidly syphoned out and a further batch of the oxide is added.

At the carbon anodes, oxides of carbon are formed. Simplistically this may be written

$$2O^{2-} \text{ (solution)} \rightarrow O_2(g) + 4e^-$$

followed by

$$C(s) + O_2(g) \rightarrow CO_2(g)$$

The dissolved oxide is more likely to ionise as Al^{3+} and AlO_3^{3-}, and the anode process may be more complicated than mere liberation of oxygen and its subsequent attack on the anode. For GCE Advanced, the simplistic approach is adequate; you should realise that the anodes have to be moved continuously both to allow for their destruction and to allow for the changing depth of the aluminium between removal of batches of the metal. To prevent the need for

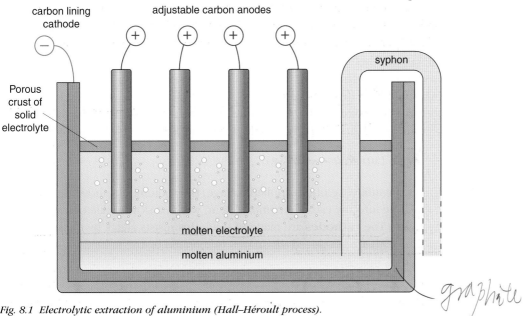

Fig. 8.1 Electrolytic extraction of aluminium (Hall–Héroult process).

stopping the process when new anodes are required, in many cells the anodes are continuously fabricated by baking a carbon paste (using the cell heat).

Between additions of aluminium oxide, the surface of the electrolyte solidifies, so the aluminium oxide can be added to pre-heat it and the surface is broken as soon as the oxide is required.

The great electrical demand of an aluminium plant puts serious limits on its location. Cheap and plentiful electricity e.g. hydroelectricity is desirable.

The high cost of aluminium makes recycling economically effective. This is helped by the widespread use of aluminium for disposable containers, the ease with which it can be separated from the other common container-metal, iron, and its low melting temperature. However, as with some other recycled materials the quality may render it unsuitable for critical uses such as the manufacture of aircraft alloy.

The electrolysis of sodium chloride

The electrolysis of sodium chloride can be used to manufacture the elements (Downs Process: electrolysis of molten sodium chloride) or can be used to **manufacture sodium hydroxide**, hydrogen and **chlorine** by electrolysis of concentrated aqueous sodium chloride (brine). Some knowledge of the latter process is required by the Edexcel specification.

It is sufficient to know that the cell is divided in such a way as to *allow* the free passage of ions (under the influence of the electric field) but to *prevent* the mixing of the products because chlorine reacts with sodium hydroxide. In the most common methods the electrode compartments are separated either by a mechanically supported membrane or by a porous (asbestos) division known as a **diaphragm**. The Edexcel specification requires only knowledge of the membrane process.

Before the brine (saturated aqueous sodium chloride) is electrolysed it is treated with a little sodium hydroxide (or carbonate) to remove traces of iron and calcium salts, etc., by precipitation and filtration. The undesirable cations might interfere with the ion-exchange action of the membrane and should they pass through it they would form an unpleasant precipitate in the sodium hydroxide. The **anodes** are made of graphite or titanium, which resist attack by the liberated chlorine. The brine flows into the anode compartment where the chlorine is liberated, piped off and used (see below) or stored:

$$2Cl^-(aq) - 2e^- \rightarrow Cl_2(g)$$

Under the influence of the applied voltage, the sodium ions move through the membrane into the cathode compartment, where a slow trickle of water is provided. The **cathodes** are made of steel; this resists attack by the sodium hydroxide that is produced. Hydrogen is liberated by the reduction of water:

$$2H_2O(l) + 2e^- \rightarrow 2OH^-(aq) + H_2(g)$$

The sodium ions and hydroxide ions constitute a solution of sodium hydroxide. (They do not *form* sodium hydroxide, they *are* sodium hydroxide). The hydrogen is piped away for storage and the sodium hydroxide solution is either heated to evaporate the water or used (see below).

Sodium hydroxide is widely used, e.g. in the purification of bauxite and in the manufacture of soap and bleaches. Chlorine is widely used in the manufacture of PVC, poly(vinyl chloride) and other chlorinated polymers, insecticides and herbicides (see p. 52) and (indirectly) hydrochloric acid.

The essential difference between diaphragm and membrane cells

You are not required to know this for the Edexcel specification but you may wish to know a little about the topic. The obsolescent diaphragm cell relied on the presence of a porous sheet of asbestos to prevent gross mixing of the contents of the anode and cathode compartments. Water could pass through and the positive sodium ions were driven one way by the potential gradient (from anode to cathode); the chloride ions tended to stay in the anode compartment and move towards the anode. Nevertheless, some chloride ions tended to be carried through the diaphragm with the moving water or dragged through by the attraction of the oppositely charged sodium ions. The product, sodium hydroxide, was always contaminated with sodium chloride, so much so, that during the heating and evaporation of the sodium hydroxide solution some of the less soluble sodium chloride was filtered out as it started the process of crystallisation. If sodium hydroxide was required to be free of chloride it had to be made by the use of the mercury cell (which operated on a different principle). The use of the mercury cell has now been discontinued in most countries because of the unacceptable environmental problems associated with escaped mercury and its compounds.

A modern ion-exchange membrane does not allow the free passage of both sodium and chloride ions; it is only permeable to the sodium ion. A typical membrane (which has to be mechanically supported to prevent rupture) consists of PTFE, poly(tetrafluoroethene) $(C_2F_4)_n$, modified by the inclusion of sulphonic acid groups, $-SO_3^-H^+$ (instead of some of the F atoms). The sulphonic acid groups serve two purposes: (i) being ionic, they attract the (polar) water molecules into the otherwise waterproof PTFE to form a continuum of 'water of hydration', thus allowing entry of the (hydrated) sodium ion and (ii) the proton, H^+, can be replaced (ion exchange) by the Na^+ ion but not by the Cl^- ion. The sodium ions can jump from one sulphonic acid group to the next under the influence of the electric field until they pass into the cathode compartment. Unlike the diaphragm, there are no vast holes, caverns and passages (viewed on a molecular scale) in the membrane and mechanical passage of the brine through it is impossible. The cathode product is thus pure sodium hydroxide and hydrogen.

Manufacture of bleach

The cheapest and commonest household bleach (typically supermarket 'own brand') is a solution of sodium chlorate(I) (sodium hypochlorite). This is made by the reaction of chlorine and aqueous sodium hydroxide, which are direct products of the above process:

$$Cl_2(g) + 2OH^-(aq) \rightleftharpoons ClO^-(aq) + Cl^-(aq) + H_2O(l)$$

It is important that the sodium hydroxide solution is cooled before the reaction since chlorate(I), the active ion in the bleach, undergoes disproportionation if heated:

$$3ClO^-(aq) \rightarrow 2Cl^-(aq) + ClO_3^-(aq)$$

Questions

1   The Haber synthesis of ammonia carried out with an iron catalyst at 450°C and 200 atm pressure, involves the following equilibrium:

$$N_2(g) + 3H_2(g) \rightleftharpoons 2NH_3(g); \quad \Delta H = -92\,kJ\,mol^{-1}$$

(a) Why is the synthesis carried out at high pressure?

(b) State one advantage and one disadvantage of performing this reaction at high temperature.

(c) Suggest a reason why the nitrogen and hydrogen must be purified before use.

(d) Assuming that the cost of obtaining the nitrogen and hydrogen gases are fixed, state one factor that would increase significantly the price of ammonia.

(e) Give two commercially important chemicals manufactured from ammonia.

2   Ethanol is made industrially by the direct hydration of ethene at 300 °C and 70 atm pressure in the presence of phosphoric(V) acid as catalyst:

$$C_2H_4(g) + H_2O(g) \rightleftharpoons C_2H_5OH(g)$$

(a) Explain qualitatively why the synthesis of ethanol is carried out:

   (i) at increased pressure;

   (ii) at a high temperature in the presence of a catalyst.

(b) State and explain the effect of increasing the proportion of water in the reaction mixture of ethene and water.

3 (a) In the electrolysis of aqueous sodium chloride:

   (i) Of what materials are the electrodes made?

   (ii) What would happen if both electrodes were made of steel (iron)?

   (iii) Classify, with the reason, the anode reaction as oxidation or reduction;

   (iv) What chemical reaction would occur *in solution* if no division were present?

   (v) Why would the presence of sodium chloride in sodium hydroxide be unlikely to cause a problem in soap manufacture? You may need to find out about this topic in the library.

(b) Read the passage (p. 84) on the difference between diaphragms and membranes:

   (i) Give the structure of PTFE;

   (ii) Give one reason why PTFE might be a suitable material to use as the basis for membrane manufacture;

   (iii) Na^+ ions pass through the membrane by changing places with the H^+ ions of the sulphonic acid group. Why cannot Cl^- ions pass through by exchanging with the much more abundant F atoms in the membrane?

Assessment questions

The following questions are all taken from Edexcel Unit 2 tests for the years 2001 and 2002.

1 (a) (i) Explain the term homologous series. **[2]**

    (ii) To which homologous series does ethene, C_2H_4, belong? **[1]**

(b) Draw the full structural formulae, showing all the bonds, for each of the following:

    (i) the organic product of the reaction of ethene, C_2H_4, with aqueous potassium manganate(VII) and sulphuric acid **[2]**

    (ii) 3,4-dimethylhex-2-ene **[2]**

    (iii) a repeating unit of poly(propene). **[2]**

(c) Ethene reacts with hydrogen chloride gas to form C_2H_5Cl.

    (i) What type of reaction is this? **[2]**

    (ii) Give the systematic name for C_2H_5Cl. **[1]**

(Total 12 marks)
(June 2001)

2 The reaction between sulphur dioxide and oxygen is a dynamic equilibrium.

$$2SO_2 + O_2 \rightleftharpoons 2SO_3 \quad \Delta H = -196 \text{ kJ mol}^{-1}$$

(a) Explain what is meant by dynamic equilibrium. **[2]**

(b) Copy and complete the Table, stating the effect on this reaction of increasing the temperature and of increasing the pressure. **[3]**

	Effect on the rate of reaction	Effect on the position of equilibrium
Increasing the temperature	Increases	←
Increasing the pressure	*increases*	→

(c) This reaction is one of the steps in the industrial production of sulphuric acid. The normal operating conditions are a temperature of 450 °C and a pressure of 2 atmospheres and the use of a catalyst.

Justify these conditions:
    (i) a temperature of 450 °C **[3]**

    (ii) a pressure of 2 atmospheres **[2]**

    (iii) a catalyst. **[1]**

(d) Give the name of the catalyst used. **[1]**

(e) Give one large scale use of sulphuric acid. **[1]**

(Total 13 marks)
(June 2001)

3 The enthalpy change of combustion of two fuels is listed below:

Fuel	Enthalpy of combustion/ kJ mol^{-1}
Hydrogen, H_2	−280
Octane, C_8H_{18}	−5510

(a) Calculate the enthalpy change per unit mass for each of the fuels. **[3]**

(b) Suggest, giving two reasons, which substance is the more useful as a fuel for motor cars. **[2]**

(c) Suggest one disadvantage of the fuel chosen in (b). **[1]**

(Total 6 marks)
(June 2001)

4 Aluminium is manufactured by a process in which purified bauxite, dissolved in molten cryolite, is electrolysed at 800 °C. Graphite electrodes and a current of 120 000 amperes are used.

(a) (i) Give the ionic equations for the reactions taking place at the electrodes. **[2]**

2Al^{3+} (l) + 6e$^-$ → 2Al (l), 3O^{2-} (l) → 1½O$_2$ (g) + 6e$^-$

    (ii) State which of these reactions is an oxidation process. **[1]**

second

87

ASSESSMENT QUESTIONS

(iii) Explain why the anodes need to be replaced frequently. **[2]**

The bauxite needs plenty of electrons to be molten which is economic.

$C(s) + O_2(g) \rightarrow CO_2(g)$

(iv) Explain why an electrolyte of pure molten bauxite is not used. $\rightarrow 2000°C$ (melting point) **[2]**

(b) The production of aluminium is expensive. *dangerous for shell*

(i) Explain why, despite this high cost, aluminium is manufactured in large quantities. *It has a large number of uses, e.g. beer cans,* **[2]**

(ii) Explain why it is worthwhile to recycle aluminium. *cooking foil* **[2]**

The cost of extraction is very high.

electrolysis is expensive recycling avoid cost step, save resources

(Total 11 marks)

(June 2001)

5 *(a)* Predict the structural formula of the organic product from the reaction of 1-bromopropane with:

(i) aqueous potassium cyanide solution **[1]**

(ii) ammonia gas. **[1]**

(b) Give details of a chemical test you could do to distinguish between 2-chlorobutane and butan-2-ol, including the expected observations with each compound. **[2]**

(c) (i) Draw the full structural formula, showing all the bonds, for the isomer of butan-2-ol that is a tertiary alcohol. **[1]**

(ii) Give details of a chemical test you could do to distinguish between butan-2-ol and its isomer drawn in (i) and the observations you would expect to make. **[4]**

(iii) Explain the chemistry involved in the test you described in part (ii) **[2]**

(Total 11 marks)

(June 2001)

6 In the preparation of the alcohol butan-2-ol, 13.7 g of 2-bromobutane was hydrolysed with 9.0 g of potassium hydroxide in aqueous solution. The following reaction occurred:

$$CH_3CHBrCH_2CH_3 + KOH \rightarrow CH_3CHOHCH_2CH_3 + KBr$$

(a) Calculate the amount (number of moles) of each reactant in the above experiment and use

your answers to state which reactant was present in excess. **[4]**

(b) Calculate the maximum possible mass of butan-2-ol which could be obtained in the above experiment. **[3]**

(c) The reaction taking place can be classified as nucleophilic substitution. Explain the term nucleophile and identify the nucleophile in the reaction. **[2]**

(d) The above experiment was repeated under identical conditions, except that 2-iodobutane was used instead of 2-bromobutane. State and explain the effect that this would have on the rate of reaction. **[2]**

(Total 11 marks)

(Jan 2002)

7 *(a)* A sample of 2-bromobutane was heated with potassium hydroxide in ethanolic solution. A reaction occurred producing a mixture of but-1-ene and but-2-ene.

(i) Write an equation for the above reaction to show the production of either but-1-ene or but-2-ene. **[1]**

(ii) State the type of reaction taking place. **[1]**

(b) Some bromine solution was shaken with a sample of but-2-ene and a reaction occurred.

(i) State what would be seen during this reaction. **[1]**

(ii) Draw the structural formula of the product of this reaction and name this product. **[2]**

(c) But-2-ene can exist as two stereoisomers.

(i) Draw the structural formulae of the two stereoisomers of but-2-ene. **[2]**

(ii) Explain why but-2-ene exists as two stereoisomers and name this type of isomerism. **[2]**

(Total 9 marks)

(Jan 2002)

8 (a) (i) Define the term standard enthalpy of formation, $\Delta H_f^{\ominus}$. **[3]**

(ii) The following table shows some values of standard enthalpy of formation:

Name	Formula	$\Delta H_f^{\ominus}$/kJ mol^{-1}
Ethene	$C_2H_4(g)$	+52.3
Hydrogen bromide	HBr(g)	−36.2
Bromoethane	$C_2H_5Br(g)$	−60.4

Use the data in the Table to calculate the standard enthalpy change for the following reaction:

$$C_2H_4(g) + HBr(g) \rightarrow C_2H_5Br(g)$$ **[2]**

(iii) State the significance of the sign of the value obtained in part (a)(ii). **[1]**

Bond	Average bond enthalpy/kJ mol^{-1}
C=C	+612
C—C	+348
C—H	+412
C—Br	+276
H—Br	+366

(b) Enthalpy changes can also be calculated using average bond enthalpy data:

Use the data in the Table above to recalculate the enthalpy change for the reaction:

$$C_2H_4(g) + HBr(g) \rightarrow C_2H_5Br(g)$$ **[3]**

(c) Suggest why the value obtained in part (b) above is likely to be less accurate than that obtained in part (a)(ii). **[2]**

(Total 11 marks)
(Jan 2002)

9 The following Table shows some properties of two different fuels:

Fuel	Hydrogen	Ethanol
Formula	H_2	C_2H_5OH
Boiling temperature/°C	−252	78
Enthalpy of combustion per gram/kJ	−143	−30
Cost per tonne/£	100	500

(a) Write equations for the complete combustion of:
(i) hydrogen
(ii) ethanol. **[4]**

(b) Hydrogen and ethanol are used as motor car fuels. Suggest the advantages and disadvantages of each of these fuels in this application. Use the Table and your answers to part (a) to help you. **[5]**

(Total 9 marks)
(Jan 2002)

10 The action of an acid with a base to give a salt is an exothermic reaction. In an experiment to determine the enthalpy of neutralisation of hydrochloric acid with sodium hydroxide, 50.0 cm^3 of 1.00 mol dm^{-3} HCl was mixed with 50.0 cm^3 of 1.10 mol dm^{-3} NaOH. The temperature rise obtained was 6.90 °C

(a) Define the term enthalpy of neutralisation. **[1]**

(b) Assuming that the density of the final solution is 1.00 g cm^{-3} and that its heat capacity is 4.18 J K^{-1}g^{-1}, calculate the heat evolved during the reaction. **[3]**

(c) 0.0500 mol of acid was neutralised in this reaction; calculate $\Delta H_{neutralisation}$ in kJ mol^{-1} **[2]**

(d) Suggest why the sodium hydroxide is used in slight excess in the experiment. **[1]**

(Total 7 marks)
(May 2002)

ASSESSMENT QUESTIONS

11 *(a)* (i) State two factors other than a change in temperature or the use of a catalyst that influence the rate of a chemical reaction. **[2]**

    (ii) For one of the factors you have chosen explain the effect on the rate. **[2]**

(b) The Maxwell–Boltzmann distribution of molecular energies at a given temperature T_1 is shown below:

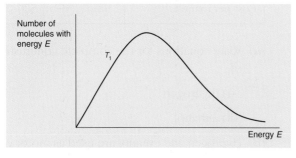

    (i) Copy the diagram and on the same axes draw a similar curve for a reaction mixture at a higher temperature T_2. **[2]**

    (ii) Place a vertical line marked E_a at a plausible value on the energy axis to represent the activation energy for a reaction. **[1]**

    (iii) Use your answers to parts *(i)* and *(ii)* to explain why an increase in temperature causes an increase in the reaction rate. **[3]**

(Total 10 marks)

(May 2002)

12 *(a)* (i) Define the term standard enthalpy of combustion. **[2]**

    (ii) The values for the standard enthalpy of combustion of graphite and carbon monoxide are given below:

	$\Delta H_c^{\ominus}$ /kJ mol^{-1}
C (graphite)	−394
CO(g)	−283

Use these data to find the standard enthalpy change of formation of carbon monoxide using a Hess's law cycle:

$$C(graphite) + \tfrac{1}{2}O_2(g) \rightarrow CO(g) \qquad \textbf{[3]}$$

    (iii) Suggest why it is not possible to find the enthalpy of carbon monoxide directly. **[1]**

    (iv) Draw an enthalpy diagram for the formation of carbon monoxide from graphite. **[1]**

(b) Natural gas consists of methane, CH_4. When methane burns completely in oxygen the reaction occurs as shown in the equation:

$$CH_4(g) + 2\,O_2(g) \rightarrow CO_2(g) + 2\,H_2O(l)$$
$$\Delta Hc = -890 \text{ kJ mol}^{-1}$$

Methane does not burn unless lit.

Use this information to explain the difference between thermodynamic and kinetic stability. **[4]**

(c) Suggest why methane is not used as a fuel for motor vehicles whereas octane is.

The following data may be useful:

$$CH_4(g): \Delta H_c = -37 \text{ kJ m}^{-1}$$

$$C_8H_{18}(l): \Delta H_c = -3530 \text{ kJ m}^{-1} \qquad \textbf{[3]}$$

(Total 14 marks)

(May 2002)

The Periodic Table of Elements

Group

Key

Atomic number
Symbol
Name
Molar mass in g mol^{-1}

Period	1	2		3	4	5	6	7	0
1	1 H Hydrogen 1								2 He Helium 4

Period 2:
| 3 Li Lithium 7 | 4 Be Beryllium 9 | | 5 B Boron 11 | 6 C Carbon 12 | 7 N Nitrogen 14 | 8 O Oxygen 16 | 9 F Fluorine 19 | 10 Ne Neon 20 |

Period 3:
| 11 Na Sodium 23 | 12 Mg Magnesium 24 | | 13 Al Aluminium 27 | 14 Si Silicon 28 | 15 P Phosphorus 31 | 16 S Sulphur 32 | 17 Cl Chlorine 35.5 | 18 Ar Argon 40 |

Period 4:
| 19 K Potassium 39 | 20 Ca Calcium 40 | 21 Sc Scandium 45 | 22 Ti Titanium 48 | 23 V Vanadium 51 | 24 Cr Chromium 52 | 25 Mn Manganese 55 | 26 Fe Iron 56 | 27 Co Cobalt 59 | 28 Ni Nickel 59 | 29 Cu Copper 63.5 | 30 Zn Zinc 65.4 | 31 Ga Gallium 70.4 | 32 Ge Germanium 73 | 33 As Arsenic 75 | 34 Se Selenium 79 | 35 Br Bromine 80 | 36 Kr Krypton 84 |

Period 5:
| 37 Rb Rubidium 85 | 38 Sr Strontium 88 | 39 Y Yttrium 89 | 40 Zr Zirconium 91 | 41 Nb Niobium 93 | 42 Mo Molybdenum 96 | 43 Tc Technetium (99) | 44 Ru Ruthenium 101 | 45 Rh Rhodium 103 | 46 Pd Palladium 106 | 47 Ag Silver 108 | 48 Cd Cadmium 112 | 49 In Indium 115 | 50 Sn Tin 119 | 51 Sb Antimony 122 | 52 Te Tellurium 128 | 53 I Iodine 127 | 54 Xe Xenon 131 |

Period 6:
| 55 Cs Caesium 133 | 56 Ba Barium 137 | 57 ▲ La Lanthanum 139 | 72 Hf Hafnium 178 | 73 Ta Tantalum 181 | 74 W Tungsten 184 | 75 Re Rhenium 186 | 76 Os Osmium 190 | 77 Ir Iridium 192 | 78 Pt Platinum 195 | 79 Au Gold 197 | 80 Hg Mercury 201 | 81 Tl Thallium 204 | 82 Pb Lead 207 | 83 Bi Bismuth 209 | 84 Po Polonium (210) | 85 At Astatine (210) | 86 Rn Radon (222) |

Period 7:
| 87 Fr Francium (223) | 88 Ra Radium (226) | 89 ▲▲ Ac Actinium (227) | 104 Rf Rutherfordium (261) | 105 Db Dubnium (262) | 106 Sg Seaborgium (263) | 107 Bh Bohrium (264) | 108 Hs Hassium (269) | 109 Mt Meitnerium (268) | 110 Uun Ununnilium (269) | 111 Uuu Unununium (272) | 112 Uub Ununbium (277) |

▲ Lanthanide elements

| 58 Ce Cerium 140 | 59 Pr Praseodymium 141 | 60 Nd Neodymium 144 | 61 Pm Promethium (147) | 62 Sm Samarium 150 | 63 Eu Europium 152 | 64 Gd Gadolinium 157 | 65 Tb Terbium 159 | 66 Dy Dysprosium 163 | 67 Ho Holmium 165 | 68 Er Erbium 167 | 69 Tm Thulium 169 | 70 Yb Ytterbium 173 | 71 Lu Lutetium 175 |

▲▲ Actinide elements

| 90 Th Thorium 232 | 91 Pa Protactinium (231) | 92 U Uranium 238 | 93 Np Neptunium (237) | 94 Pu Plutonium (242) | 95 Am Americium (243) | 96 Cm Curium (247) | 97 Bk Berkelium (245) | 98 Cf Californium (251) | 99 Es Einsteinium (254) | 100 Fm Fermium (253) | 101 Md Mendelevium (256) | 102 No Nobelium (254) | 103 Lr Lawrencium (257) |

Index

Page references in *italics* refer to a table or an illustration.

INDEX

INDEX